KT-130-082

CONTENTS

3

FOREWORD

In 2008 we celebrated 150 years since the presentation, by Charles Darwin and Alfred Russel Wallace, of their ideas on evolution to the world. Then, the first fossil human finds were only beginning to be made or recognized, and evolutionary ideas, palaeontology and archaeology were still in their infancy. Now, there is a rich and ever-growing record from Africa, Asia and Europe, and a plethora of accompanying scientific names, some of which are difficult to articulate, even for the specialists! In this book, Charles Lockwood guides us expertly through the evidence for over 20 of the fossil species that represent our ancestors or close relatives.

New finds from various regions of Africa are fleshing out the possible beginnings of the human evolutionary line, with several candidate species vying for the title of our earliest ancestor. The surprisingly early spread of humans from Africa is well illustrated by the unexpectedly rich site of Dmanisi in Georgia, dating close to 2 million years ago. Here, skulls, jaws and parts of skeletons of some primitive human fossils have been found beneath the remains of a medieval village. The small brain size and very basic stone tool kit of these early humans are surprising; many experts assumed that significant advances in intelligence and technology would have been necessary for these early humans to make the first moves out of their ancestral African homelands. Indeed, the evidence is so challenging, it has even been suggested that we rethink the idea that the genus *Homo* originated in Africa!

Fossils ranging from South Africa to Australia are revealing the origin and spread of our own species from Africa. How little we yet know about later human evolution, in regions such as Southeast Asia, has been highlighted by the discovery of a remarkable skeleton of a very primitive, human-like form on the island of Flores, Indonesia. The existence of this creature, given the name *Homo floresiensis* and nick-named the 'Hobbit', was completely unsuspected. The fact that it apparently survived into

CHARMINSTER LIBRARY

THE
HUMAN
STORY

Where we come from
and how we evolved

CHARLES LOCKWOOD
Foreword by Chris Stringer

BOURNEMOUTH
LIBRARIES
WITHDRAWN STOCK
FOR SALE
CUA
410082740

Publish   London

First published by the Natural History Museum, Cromwell Road, London SW7 5BD
© Natural History Museum, London, 2007
Reprinted with updates 2008, 2010, 2013 update by Chris Stringer, Louise Humphrey, Isabelle De Groote and Laura Buck

ISBN 13: 978 0 565 09322 8

All rights reserved. No part of this publication may be transmitted in any form or by any means without prior permission from the British publisher.

A catalogue record for this book is available from the British Library.

Edited by Karin Fancett
Designed by Mercer Design
Reproduction by Saxon Photo Litho
Printing by C&C Offset

Front cover: Only about 2000 generations have passed since *Homo neanderthalensis*, Neanderthal man, became extinct. © Michael Long/NHMPL
Back cover: The incredible skeleton of *Australopithecus afarensis*, Lucy, uncovered after 3 million years. © Natural History Museum, London
Spine: Skull of *Homo heidelbergensis*. © Natural History Museum, London

the last 50,000 years means that modern humans, dispersing through the region to Australia, may have encountered these strange relatives. The evidence is so challenging to conventional thinking that some experts have questioned its reality, insisting that the unusual features of the material are the products of abnormality rather than evolution.

As Charles explains, despite all the new finds and advances in study techniques, there are still many fascinating puzzles about our evolutionary origins. The exact nature of the last common ancestor we shared with chimpanzees remains uncertain, as is the date at which it lived, and the environment in which it originated. There are also many different ideas about why our ancestors began the fundamental human adaptation of walking upright on two legs, but we are still far from knowing whether any of them are correct. And although most experts believe the event happened in Africa prior to 2 million years ago, we do not really know when, where and why the first members of the genus *Homo* evolved. Equally, for *Homo sapiens*, what processes were involved in the origin of our species, and how did we come to replace those other surviving species, such as the Neanderthals and the strange 'Hobbit' from Flores? Further finds will help us to close many of the gaps in our knowledge, but in the meantime, Charles's book provides a readable, up-to-date, well-illustrated and user-friendly summary of the evidence as it stands today.

Sadly, Charles died in a traffic accident in London on 14 July 2008. He was already an established figure in research on human evolution and was about to take up the position of Director of the Institute for Human Evolution in South Africa. He is sorely missed.

Professor Chris Stringer
Research Leader in Human Origins and Director of
Ancient Britain in its European Context Project
Department of Palaeontology
Natural History Museum
London

INTRODUCTION

Here is a tongue twister for you: *Australopithecus anamensis* was an ancestor of *Australopithecus afarensis*. It is difficult to say even once. But this tongue twister means something – it refers to two of our earliest ancestors who existed long before written history, before humans spread throughout the world, and even before human-like creatures made their first tools from stone. The record of human evolution extends back more than 6 million years, to the point when our lineage split from the chimpanzee lineage.

We are lucky to have an excellent record of bones and stone tools that tell us where, when and how humans evolved. The record is so extensive that scientists have now named more than 20 species in this 6-million-year span. Each of these species was a group of individuals who were genetically closely related and who bred with one another. Humans are a species – *Homo sapiens* – and so are chimpanzees (*Pan troglodytes*) and gorillas (*Gorilla gorilla*). When we talk about human evolution, it is important to name species so that we can communicate about the groups of fossils associated with that species, determine how they behaved, and find out how they are related to us.

All of the species that are more closely related to humans than to any of the living apes are referred to as 'hominins', members of the taxonomic group Hominini. In this book, each of the hominin species is described roughly in order of when it lived, grouped into five main sections: earliest hominins, *Australopithecus*, *Paranthropus* (or robust *Australopithecus*), early species in the genus *Homo* and finally the later *Homo* species – including us, *Homo sapiens*. The complete human story emerges through the book, but it does not have to be read in order. If you want to know, for example, what *Homo habilis* was, you can turn to that section of the book and find out. Here, we introduce the characters that play out the human story, a story that involves a large and diverse cast.

Evidence from bones

Skeletons of three hominin species: *Homo erectus*, *Australopithecus afarensis* and *Homo sapiens* (from left to right).

EARLIEST HOMININS

Genetic comparisons of living humans and apes tell us that our closest relative is the chimpanzee, and that we split from the chimpanzee lineage between 6 and 8 million years ago. For a long time in the study of human evolution we had no fossils that could tell us about the earliest hominins. Recently that has changed. Here, the species have all been described as ancestors to later hominins, but scientists are not entirely certain about the role each species plays in human evolution. The different names show the uncertainty – there is limited evidence with which to compare the species directly and determine how many distinct groups are represented. It will take time to know exactly how to interpret the fragmentary fossils, and the names may change too. Scientists may decide to lump the various discoveries into a smaller number of species. Despite this we can still ask, 'What were the earliest hominins like?'

Searching for the most ancient hominins

Palaeontologists working in the Djurab desert of northern Chad, where *Sahelanthropus* was found. These sites are between 6 and 7 million years old.

Identifying the first hominins

Before we even look at the fossil record, we can make predictions about what the first hominins would look like. In the ancestor that we share with apes, we would expect to see anatomical features and behaviours shared by chimpanzees, gorillas and orang-utans – the living great apes. Starting from that ancestral point, the first hominins must have evolved at least one feature that we see in modern humans. They would not be expected to look just like us, but would show at least one characteristic that places them on the human lineage.

In looking for these features scientists focus on two major areas: anatomy related to walking upright on two legs – bipedalism – and the size and shape of the canine and first premolar teeth. The fossil record shows that some other important human characteristics, such as large brains, come much later in time, so these features are less important in identifying the first hominins. Humans are the only primates to routinely walk bipedally, and therefore bipedal walking is an obvious feature to look for in comparing hominins to other species. Many other primates walk bipedally for short periods or stand upright to obtain food, but no other species moves this way as its principal way of getting around.

Differences between humans and apes in their front teeth appear to be more subtle, but they are just as important as bipedalism in identifying the first hominins. Humans have smaller canine teeth than apes and monkeys. Because our canines are small, the lower first premolar has changed as well. In apes and monkeys, this

Standing up

Unlike apes and monkeys – shown here by a chimpanzee (middle) and a baboon (right) – humans (left) are adapted to walking upright. This impacts our skeleton in many ways. For example, curvature of the lower back allows us to keep our trunk over our hips without much effort.

tooth has a single point (cusp) and a long 'honing' surface – it sharpens the fang-like upper canine as the jaws interlock and the two teeth wear against each other. In us, it has two cusps and is completely different in shape.

Going from a large, fang-like canine to a tiny one clearly requires intermediate stages, and for the most part the fossil record shows this transition well. One of the reasons that the earliest hominins are called hominins is because they show a number of small changes in the canines and premolars, and it appears that their lower first premolar is no longer used for honing. Of course, the canines of early hominins are still much larger than ours! It is only in comparison to the apes in general that their canines are called 'small'. Deciding whether fossils are hominin depends on the results of detailed comparisons of key features such as these.

Hominin anatomy

The skull and upper jaw of *Australopithecus afarensis* (see p. 28) are shown here between a chimpanzee and a human. Chimpanzees have large canine teeth for use as weapons. Canines are smaller in fossil hominins and humans.

Sahelanthropus tchadensis

When the discovery of *Sahelanthropus tchadensis* (Sahel man of Chad) was announced in 2002, Bernard Wood remarked in the journal *Nature* that it "fundamentally change[s] the way we reconstruct the tree of life". Palaeoanthropologists such as Wood were astounded at the location, the date and the appearance of this new species. The site in Chad, in central Africa, is far to the west of most early hominin sites, and animals found with *Sahelanthropus* suggest it is around 7 million years old. This makes it the oldest species yet found that may be a hominin. When you first look at the face of the skull, you see a modern-looking form with a hulking browridge, not unlike *Homo erectus* – a much more recent species described later in the book. But the rest of the *Sahelanthropus* skull is decidedly more ape-like than other hominins. Hence, the conundrum for scientists: where does 'Sahel man' belong in the human family tree?

Oldest species

Nicknamed 'Toumai' – or 'hope of life' – this remarkable skull is between 6 and 7 million years old and is the most complete specimen of *Sahelanthropus*.

The first biped?

For *Sahelanthropus*, no bones of legs and feet have been found to determine if it was bipedal. However, the position and orientation of its foramen magnum – the hole in the base of the skull through which the spinal cord passes – suggests a posture similar to bipeds. It is horizontally oriented and relatively far forward, just what is needed if you are walking upright and looking straight ahead.

Sahelanthropus also may have had small canines and a hominin type of premolar – one that does not sharpen the upper canine. These changes in the front teeth are just as important as bipedalism in identifying the first hominins.

Lifestyle

It is likely that *Sahelanthropus* ate mainly fruit, with smaller amounts of a variety of other foods as well. But we need more evidence to reconstruct its diet in detail. If *Sahelanthropus* was bipedal, that would have affected the way it moved around the landscape and the way it obtained food, but apparently the types of food were not that different from the range of foods eaten by modern apes.

An important but unresolved question with *Sahelanthropus* is the exact nature of its habitat. The key specimens were found with a variety of habitats close to each other, ranging from desert conditions to the thick forests surrounding rivers. It is difficult to tie the species to one of these environments, although it is intriguing that from this early time period, open habitats such as grasslands are found nearby hominins, and these contrast with the forested habitats of modern apes.

Orrorin tugenensis

Sahelanthropus may be the oldest species discussed in this book, but the first discovery that broke the 5-million-year mark in human evolution is a species from Kenya. It was called *Orrorin tugenensis*, orrorin means 'original man' in the local language of the Tugen Hills, the region where it was found. Again, its date of 5.8 million years is in the general time range that geneticists consider to be the time of origin of the human lineage. The date is reliable, as it comes from absolute dating of volcanic layers (lava flows) above and below the fossils.

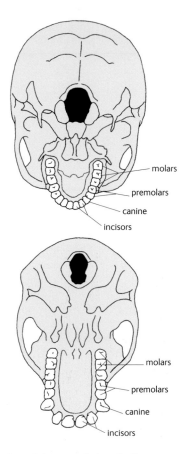

Bipedalism and the skull

The large hole in the base of the skull is where the spinal cord enters. In humans (top), it is further forward than in the apes (bottom), because of our upright posture.

Tugen Hills, Kenya

Teeth, jaws and limb bones of *Orrorin* were discovered in this region in the year 2000.

Legging it

Orrorin is unusual as it is known mainly from parts of its skeleton other than its skull. There are fragments of jaws and teeth, but the main evidence about *Orrorin*'s lifestyle comes from two thighbones, especially the parts that connect with the hip. These thighbones (femurs) may be the oldest direct evidence for hominins walking on two legs.

Walking upright requires a substantially different hip joint to that seen in apes. In humans, the joint is stabilised by ligaments running across the joint between the pelvis and the femur. The marks left by these ligaments are clearly visible on most human femurs but not in most apes. When such marks were observed on *Orrorin* fossils, part of the case for bipedalism was made.

Another key difference between humans and apes is the structure of bone inside this part of the femur. With each step, humans bear all of their weight through this one joint, and nearly all the stress is placed on the bottom edge of the femoral neck – the narrow structure between the long part of the femur and its round 'head' (the ball part of the ball-and-socket hip joint). As a result, bone is denser in the bottom part of the human femoral neck, while in apes the hard bone is spread evenly around the edges.

In well-accepted bipeds such as *Australopithecus afarensis* (see pp. 22–30), the distribution of this hard bone is often cited as good evidence for bipedalism. It can be seen either in naturally

Signs of walking

Later bipeds, such as *Australopithecus afarensis* (middle), have a characteristic pattern of bone distribution in the neck of the femur (thighbone) that is similar to humans (right) and unlike chimps (left). Bone is more dense at the bottom edge of the femoral neck, because this area is subject to the most stress when walking upright.

broken fossils or in specimens that have been scanned with medical imaging equipment. With *Orrorin*, the same pattern is more subtle, resulting in some disagreement about whether this is valid evidence for bipedalism.

A different picture from the teeth

Orrorin is different from *Sahelanthropus* in that it has ape-like canines and premolars. The primary evidence for it being a hominin, however, is in the leg bones. Its discoverers also pointed out that *Orrorin* has thick tooth enamel, like humans. However, since some other apes, such as orangutans, have thick enamel as well, interpreting this feature in *Orrorin* is not straightforward. As you can see, different kinds of evidence have been used to justify each of these species – *Sahelanthropus* and *Orrorin* – as a hominin, and it may be that only one of them (or neither) is actually part of our ancestry.

The dating game

To work out which is the first hominin, we need to know how old these discoveries are in geological terms. Dates for *Orrorin* and *Ardipithecus* fossils are reliable, because they come from 'absolute' dating techniques. The most common techniques are radio-isotopic methods, making use of the process of radioactive decay and the fact that this process behaves more or less like a clock.

We do not always know the dates so well, because these clocks have to be set by particular conditions. Eastern African hominin sites are a perfect setting, because volcanic activity was common through the period during which fossils were

deposited. Volcanic eruptions 'reset' a process of decay for an isotope of potassium to argon, because the high level of heat eliminates argon gas, and the argon begins to accumulate from a baseline of zero. Because of this, volcanic layers such as ash or lava can be dated with precision. Fossils themselves are not dated with this method, but when layers above and below them are dated, we know the age of the fossil that lies between them.

Many early hominin sites are studied with potassium-based methods. Much more recently, at sites younger than 40,000 years old, the carbon-dating method is used. The death of an organism essentially starts a clock

The 'Sidi Hakoma tuff' is a volcanic ash layer in Ethiopia from an eruption 3.4 million years ago. In some places these layers reach incredible thicknesses of 5–10 m (16.4–32.8 ft), because ash was piled up in a river channel.

The motion of the river current is immortalized in the rock.

in which the radioactive carbon-14 isotope begins to decay within the fossil, and it makes up less and less of the total carbon through time.

Where radioisotopic methods cannot be applied to fossil sites, other methods are useful. The most common is to use the types of animals that are discovered to place the site at the most logical time period, based on the dates known for those animals at sites with good calibration. This is called biostratigraphy or biochronology, and the date for *Sahelanthropus* was determined this way.

Ardipithecus ramidus and *Ardipithecus kadabba*

The *Ardipithecus* genus is a group that lived between 5.8 and 4.3 million years ago. In 1994, the first *Ardipithecus* fossils were announced by a team, led by Tim White, working in the Middle Awash region of Ethiopia. Since then, similar fossils have been found at other sites in Ethiopia. As with *Sahelanthropus* and *Orrorin*, we see a shift in the anatomy of teeth that indicates these could belong to the hominin group. A relatively complete female skeleton has also revealed that *A. ramidus* was bipedal for at least some of the time. *Ardipithecus* uses the word 'ardi', meaning 'ground' in the Afar language of people who live in the area where these fossils were found. The species names refer to the view of this group as an ancestor for later hominins – in Afar, 'ramid' means 'root' and 'kadabba' means a family ancestor.

Ethiopia

The Gona site, one of the places in Ethiopia where *Ardipithecus* fossils have been found.

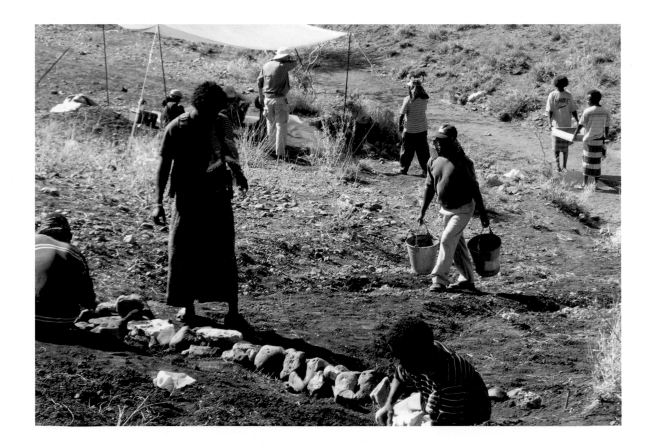

Yet another biped?

From initial reports, the most definitive evidence for calling *Ardipithecus* a hominin is the size and shape of the teeth. Although we would see *Ardipithecus* canines as large and 'ape-like', they have begun to function more like those found in humans. The first premolar does not hone the upper canine when teeth come together. Instead, as in humans, the canines wear on their tips rather than the back edge.

Sahelanthropus and *Orrorin* illustrate the key debates about early bipedalism. Can bipedalism be determined from the skull? Could the earliest hominins be as efficient at bipedalism as humans are today?

In 2009 researchers revealed the oldest known skeleton of *Ardipithecus ramidus*. It was actually discovered in 1992 but took a long time to study because it was very fragile. The skeleton is not the oldest potential hominin, but it is the most complete from this early period. Its anatomy reveals a mixture of ape- and human-like features that indicate an 'intermediate' form of upright walking. Scientists had previously expected the skeleton of our earliest ancestor to resemble a knuckle-walking chimpanzee with some human features. Surprisingly, the *A. ramidus* skeleton does not show this kind of morphology, but instead shows a combination of features seen in earlier apes and humans.

The lower face is less protruding and more vertical than that of a chimpanzee and the cranial base is short, like it is in later bipeds. The foot looks primitive, with an opposable big toe for grasping onto branches, but the morphology of the toe bones would have made the foot more rigid, aiding bipedal walking. The pelvis required extensive reconstruction because of its fragmentary condition but it appears shorter and broader than the pelvis of apes, more like that of humans. Combined with a human-like curved spine, the foot and pelvis morphology imply that *A. ramidus* was able to walk bipedally, but still probably spent a lot of time in the trees. This conclusion has led some scientists to question the status of *A. ramidus* in the human lineage, but until more evidence is found from these early stages of human evolution, the debate remains open.

Tiny clues

Early hominins can be recognised even from small pieces of bone if they contain the right information. This piece contains a first premolar tooth, which shows the initial signs of hominin morphology in *Ardipithecus ramidus*.

Early diets

When we think about how early hominins lived, we want to know whether modern apes provide a good picture or whether our earliest ancestors were different from living species. We also compare them with what came later to see whether the hominins had a common set of behaviours. Diet can be reconstructed from tooth anatomy, thickness of tooth enamel, how the tooth is worn down from chewing, and in some cases the chemical make-up of teeth and bones. With *Ardipithecus*, a key feature is relatively thin tooth enamel compared to later hominins. This suggests that *Ardipithecus* did not eat the hard foods that were more common in *Australopithecus* diets later in time. *Ardipithecus* probably ate many of the foods that chimpanzees eat today, especially fruit but also a variety of other foods (e.g. chimpanzees eat termites and ants and even hunt small monkeys). In its diet, *Ardipithecus* was probably more similar to *Sahelanthropus* than to *Orrorin*, which had thicker tooth enamel.

Cousins

Chimpanzees are our closest living relatives; we share 98 percent of our DNA with them.

Are hominins 'apes'?

The apes are our closest living relatives. The bigger species – chimpanzees, gorillas and orangutans – are referred to as great apes. Within the great apes, chimpanzees are most closely related to humans, and gorillas are slightly more distant from us. Darwin recognised the close relationship of humans to African apes and predicted that the earliest human fossils would be found in Africa. He was right.

Because great apes are closely related to us, they are usually classified within the human family, Hominidae. This book focuses on a narrower group, Hominini, which means specifically the fossil species that are more closely related to us than chimpanzees are. (In other places, you may see the name 'hominid' used in the same narrow sense that this book uses 'hominin' – the difference is just in the names and reflects an older taxonomy in which apes were excluded from the human family.)

It is easy to make the mistake of thinking that the apes are a uniform group, when in reality great differences exist among them. Chimpanzees are fascinating to us because of their complex social relationships and their tendency to show many 'human' behaviours. Gorillas are much bigger in size, and they eat leaves more frequently than the other apes, who mostly eat fruit. Orang-utans and gibbons (gibbons are so-called 'lesser apes') each have their own unique ways of moving around through trees, and have exceptionally long arms.

Despite differences like these, apes share a variety of features – for example, all apes lack tails, and so do humans. Apes also have relatively larger brains than most other primates. They have projecting faces and long canine teeth as well, but in these respects they are like many other primates.

How, then, do we think of the earliest hominins – the earliest human ancestors? Early hominins are often described as 'bipedal apes' because they were ape-like in the general appearance of their skulls and their relative brain size. At some point in human evolution, we cross a threshold where we begin to think of our ancestors as more like us, showing more of the distinctive features that arise during human evolution. It is worth keeping in mind that this is a subjective decision – what is ape, and what is human? There is no definite line between these categories, and as we go through the variety of hominin species, the question becomes more and more difficult to answer.

AUSTRALOPITHECUS

The discovery of *Australopithecus* in the 1920s was a key event in understanding human evolution. The full story is told in the section on *Australopithecus africanus* (see pp. 36–43). *Australopithecus* species were small-brained, bipedal human relatives. They show us that the defining hominin characteristic is bipedal walking rather than large brains or toolmaking. As well as being bipedal, *Australopithecus* species had thicker tooth enamel than the great apes, and modified canines and premolars reflecting a change in the function of these teeth. They are thus 'hominin' in all these respects. From initial comparisons, it seems that these species also behaved differently from the earliest hominins described in the previous section, so *Australopithecus* ushers in a new phase of human evolution.

Layers upon layers

The site of Hadar in Ethiopia records nearly 500,000 years in the life of *Australopithecus afarensis*. Field workers from the surrounding Afar region, such as Hamadu Mohamed Ware, play a crucial role in finding the fossil evidence.

Australopithecus afarensis

If you were to go back just over 3 million years in time and walk through the woodlands and floodplains of northeastern Ethiopia, you would see a vast array of wildlife that included a familiar creature, a hominin called *Australopithecus afarensis*. This species is familiar to us now because one of its members happened to die near a stream that quickly buried the skeleton. After millions of years of further burial and then erosion of the land surface, she was uncovered again and exposed to the elements. In 1974, Donald Johanson and Tom Gray, part of a team of researchers who had discovered the fossil abundance of the Hadar region of Ethiopia a few years before, came across first a fragment of elbow joint, then part of the skull, and eventually much of the skeleton we have come to call Lucy.

Hadar is now a desert surrounding the Awash River, a remnant of the river system that made the region more lush at the time Lucy was alive. The circumstances necessary for fossil burial and then discovery are so perfect at Hadar that it is a veritable graveyard for animals from that time. This applies to *A. afarensis*, too, and the discoveries go well beyond what is known of Lucy herself. Today, the species is known from two nearly complete skulls and a large number of jaws and teeth, as well as a comprehensive collection of limb bones.

By looking at the shape and structure of the remains, especially the limb bones and hips, scientists can work out how hominins began to walk on two legs. It is crucial to have partial or complete skeletons to tie together all of the more fragmentary pieces that are found. This is what makes Lucy so special. When she and the other *A. afarensis* specimens were described, it transformed our understanding of human evolution in general and human walking in particular.

Hips
Humans (bottom) have short, broad hips compared with apes (top), and Lucy (middle) has a similar shape.

Lucy

Lucy is deservedly famous, as her discovery revolutionised ways of thinking about early hominins, especially the way they walked.

Middle aged

When named in 1978, *A. afarensis* was the oldest known hominin species. Many more discoveries have been made since then, and *A. afarensis* is no longer the oldest hominin – in fact this species now sits approximately in the middle of human evolutionary history. At 3.0 to 3.6 million years old, *A. afarensis* comes long after species that may be the first hominins (6 to 7 million years ago) and before the origin of our own genus of hominins, the genus *Homo*, at approximately 2–2.5 million years ago.

What's in a name?

Australopithecus means 'southern ape'. The name was coined by Raymond Dart for *Australopithecus africanus* in 1924, so since then species that are broadly similar to *A. africanus* have been put into this group. When *A. afarensis* was discovered, scientists decided it was a different species from *A. africanus* but that it still belonged to *Australopithecus*.

Fossil sites

Major hominin sites in eastern and central Africa.

Afarensis means 'of the Afar region', the region of northeastern Ethiopia where the largest sample of *A. afarensis* fossils has been found. *Australopithecus afarensis* fossils have also been discovered

The Laetoli trackway in Tanzania is a
unique series of footprints made by
three hominins walking across ash
from a recent volcanic eruption at
about 3.5 million years ago.

in Tanzania (at a site called Laetoli), as well as in Kenya and
other sites in Ethiopia. Somewhat further away, a few specimens
from Chad may belong to *A. afarensis* as well (see box on
Australopithecus bahrelghazali, p. 34).

First steps

Lucy is not the only extraordinary find associated with *A. afarensis*.
The Laetoli site in Tanzania is home to a footprint trail 3.5 million
years old that is probably a trackway of *A. afarensis* individuals,
because of teeth and jaws found at the same site that resemble
those from Hadar in Ethiopia. The footprints, in combination with
the information from the Lucy skeleton and other specimens,
illustrate in dramatic fashion that hominins at this time were
bipedal.

What is it about the skeleton that indicates bipedalism? As one
might imagine, the hips and legs are the key evidence of walking
upright. Lucy's pelvis is shorter and broader than that of apes, and
it is similar to humans in that respect. The knee joint is angled, as
in humans, so that the bottom of each thighbone (the femur) is
directed towards the middle of the body. This means that, when
each step is taken, the centre of mass passes more directly over
the lower leg. Foot bones, in combination with the footprints,
show that the heel bone was pronounced, and the foot may have
formed a supporting arch in the same way that human feet do.

Habitats

The habitat of *Australopithecus afarensis* was probably a mix of wooded areas along with more open territory.

Studies of *A. afarensis* have often focused on bipedalism, mainly because the sample of bones from the skeleton is more complete than for most other early hominins. Everyone agrees that *A. afarensis* was indeed bipedal, as opposed to quadrupedal (like most monkeys) or suspensory (like orang-utans and gibbons). *Australopithecus afarensis* did not run along the ground on all fours, and it did not have extremely long arms for climbing and swinging. There is also no evidence that it was a knuckle-walker, the particular way that chimpanzees and gorillas walk on all fours by using the knuckles of their hands instead of their palms.

When on the ground, *A. afarensis* walked upright, not exactly with the same mechanics as humans today, but in a way that was far more similar to us than to modern apes. Differences in the exact pattern of walking are suggested by details of the pelvis and lower limb. For example, the blades of Lucy's hip bones flare out further to the side than they do in humans, and they do not curve around to the front as much. Muscles would have been in a slightly different position, leading to a different way of walking.

Ground or tree-dweller?

We can reconstruct how we think the species behaved from the skeleton. But there is some controversy over how much time *A. afarensis* spent on the ground, as opposed to in the trees. Arboreal behaviour is a major part of most primates' lives, and even primates that appear highly adapted for quadrupedalism on the ground retain general features of the skeleton that allow them to climb trees and obtain food or take refuge there.

Australopithecus afarensis was a chimpanzee-sized animal, smaller than modern humans, and many aspects of its skeleton fit with the idea that this species still used trees as a major food resource and/or refuge for safety and sleeping. In other words, it was bipedal when on the ground but was able to climb trees more effectively than modern humans. Its relatively long and powerful arms would be used for hauling itself up, and its relatively long feet, and the curved bones of its toes and fingers, would have helped in grasping branches.

How they lived

Despite being bipedal, species such as *A. afarensis* are often referred to as 'ape-like', with relatively small brains compared to modern humans. Also, when we look at the projecting face and large jaw of *A. afarensis*, we see essentially the face of an ape, but that of an upright-walking ape.

Although it had an ape-like appearance, *A. afarensis* represents a way of life that helps differentiate early hominins from their ancestors. Early *Australopithecus* species were 'megadont', meaning that their teeth and particularly their back teeth were enlarged compared to chimpanzees. The back teeth do not show the prominent crests seen in gorillas, which also have large teeth but use them to slice the leaves that form a big part of the gorilla diet. The jaws and teeth of *A. afarensis* were not specifically adapted for eating one type of food. Microscopic wear on the teeth – basically the marks made by food items when they are chewed – show that they were used to grind large quantities of tough food such as grass and leaves. The relatively flat teeth with thick enamel may have been used in the manner of a pestle and mortar.

Changing teeth

A set of upper teeth from *Australopithecus afarensis*, showing a more oval shape to the tooth row than seen in chimpanzees or gorillas, with smaller canines.

A baby girl

New fossils of *A. afarensis* continue to answer some questions as well as raise others. In 2006, after five years of intensive cleaning and preparation, a team led by Zeresenay Alemseged announced another skeleton of *A. afarensis*, but this one was of a young child, roughly 3 years old. It is from a site called Dikika, which lies across the river from Hadar. Much of the Dikika skeleton is present, including some rarely preserved elements such as the shoulder blade (the scapula), and one of the few hyoid bones from the entire hominin fossil record. The hyoid bone is in the neck below the base of the tongue, and it gives information about the function of the voice box (larynx). Both of these structures are relatively ape-like in the baby skeleton, but even at its young age many of the typical features of *A. afarensis* were present. How do they know it is a girl? Although it is impossible to be sure, this conclusion comes from the size of the adult canine. Because she was so young, she still had her baby teeth in, but the adult canines were forming inside the jaws. In apes, canines in males and females are very different in size but in humans they are much more similar. *Australopithecus afarensis* still showed some variation in canine size between the sexes, and the Dikika baby's adult canines are at the low end of the species range of size; she was probably a female.

Man or monkey

Australopithecus afarensis lived 3 to 4 million years ago. They walked upright like modern humans but continued to climb trees like apes.

Baby skeleton

Nicknamed 'Selam', this child is 3.3 million years old and one of the most complete juvenile skeletons of an early hominin.

Crucial evidence

Why was the discovery of *A. afarensis* so dramatic? Partly for the quality of the fossils, and partly because it held the title of 'oldest hominin' at the time. Also, the discovery of *A. afarensis* began a long debate about evolutionary relationships that has only become more complex with further discoveries. It was announced as the earliest human ancestor and a common ancestor of all later hominins. This in itself was uncontroversial, but further study of *A. afarensis* led to changing views of other species, particularly *A. africanus*, which until that point had been regarded as the early trunk species of the human family tree. These changing views raise difficult questions about how to use information from the skeleton to shed light on evolutionary relationships. Today, *A. afarensis* is still discussed as an ancestor to humans, but whether or not that was literally the case, the large sample of bones and teeth gives us crucial evidence of what hominin life was like after bipedal walking evolved but before stone tools transformed human evolution. It serves as a point of comparison for all hominins, both earlier and later species.

Australopithecus anamensis

Of the many places that have early hominin fossils, the Lake Turkana region has perhaps the greatest treasure trove. Largely situated in northwestern Kenya but stretching into Ethiopia as well, Lake Turkana is surrounded by fossil sites of all ages relevant to human evolution. One of these sites gives us the first glimpse of *Australopithecus*. Lucy and the other individuals of *Australopithecus afarensis* may be the best known of this group, but the earliest signs of *Australopithecus* belong to a different species – *Australopithecus anamensis*. The second name is derived from the Turkana word 'anam', meaning 'lake'.

Australopithecus anamensis fossils are known from two sites near Lake Turkana – Kanapoi and Allia Bay – and range in age from 3.8 to 4.2 million years ago. This age is right between the earlier *Ardipithecus* fossils and the later, well-known *A. afarensis* sample. Such a pattern leads some to conclude that human evolution at this time was a single evolving lineage.

Fossil hunters
Prospecting for fossils at the site of Kanapoi in Kenya.

Bipedal ape

Australopithecus anamensis had a human-like lower leg (tibia), but its jaws have parallel tooth rows, more like the apes than humans.

The sample of *A. anamensis* fossils is small but contains some key evidence. Perhaps the most impressive element is a tibia (shin bone) that looks remarkably like a human tibia. The thick upper part of the tibia is robust, because walking on two legs transfers all the weight through the legs – this is a clear difference from tibias of apes, which are less built up close to the knee. Also, the lower part of the *A. anamensis* tibia shows an ankle joint with human-like orientation. As shown in more detail in the section on *A. afarensis* (pp. 22–30), understanding bipedalism is a complex task. *Australopithecus anamensis* probably did not walk exactly like we do, but it appears that this species had a bipedal posture.

Hominins in their habitats

Australopithecus anamensis fossils are found in geological deposits that indicate a variety of settings, ranging from open grasslands to more wooded areas. Does this mean they lived in all of these habitats? This is an important question because we want to know how well they could adapt to different environments. Unfortunately, it is difficult to pinpoint the relationship between fossils and specific habitats. The fossil record is the result of many processes, and fossils may end up in places that do not represent where the individuals lived. We can say that there was some open country in the region where *A. anamensis* lived, but we do not know whether these hominins spent much of their time out in these open grasslands, or whether they preferred more wooded areas.

Resolving this question would help to explain behaviours such as bipedalism – it is just as important to know the ecological setting as it is to know when bipedalism evolved. Lingering in many stories of human evolution is the 'savanna hypothesis', an idea that ties bipedalism and other human behaviours to more open environments. In contrast, African apes today live mostly in forested and wooded areas, and as far as we know their immediate ancestors did as well.

Four million years ago
This is how *Australopithecus anamensis* may have looked.

33

Australopithecus bahrelghazali

The *Sahelanthropus* skull found in 2002 (see pp. 12–13) was not the first hominin to be found in Chad. That honour belongs to a jawbone discovered in 1993 by the same team that would later find *Sahelanthropus*. Michel Brunet's group classified it as a new species – *Australopithecus bahrelghazali* – the species name comes from the fact that the site is in the Bahr el Ghazal valley – to recognise that it has some unusual characteristics and is geographically distant from other hominins at the time. At 3.6 million years before present, this specimen is in the same time range as the much more well-known species *A. afarensis* in eastern Africa. Because of the limited information, and the subsequent discovery of only a few other teeth, some researchers regard this specimen as part of *A. afarensis*. Others have suggested that *A. bahrelghazali* and *Kenyanthropus platyops* (see p. 35) could belong to the same species, but additional fossils would be required to demonstrate this.

Whatever its classification, the find is a crucial reminder that our focus on eastern and southern Africa in early human evolution comes from geological circumstance, not because hominins lived only in those places. Discoveries from places like Chad help to show the true geographical range of human evolution, and it is possible that future discoveries will reveal hominins across much of Africa. To put this in the proper context, remember that we have no significant fossil record for our closest relatives, chimpanzees and gorillas. This, of course, does not mean that they did not live in Africa in the past! Sites that provide a window into the past of west and central African forest regions have simply not been found.

The first discoveries of *Ardipithecus*, one of the earliest hominins (see pp. 16–19), played a role in the rejection of simple versions of the savanna hypothesis, mainly because of the observation that some early hominin habitats were relatively wooded. Since then, the record has become more complicated, and with the discovery of *Ardipithecus*, *Sahelanthropus* and *Orrorin*, it now seems that there was a range of diverse habitats near the places where the hominin fossils were buried. Clarifying the evidence for these environments is a central long-term goal of studying human evolution.

Diversity versus lineages

Another important question in human evolution is how many species existed in the past. The number keeps growing as new discoveries are made. Sometimes, classifying species is a matter of comparing fossils from the same time period to see whether they fit patterns of variation we expect for a single species. It is like studying apes and other primates today, using their anatomy, behaviour and genetics to determine which groups form the cohesive populations that we call species. In the fossil record however, all we have are the bones. How we recognise and separate species today helps to guide us in the use of these bones to name species in the fossil record.

variation. The vast majority of specimens are teeth, but early discoveries of relatively complete skulls and a partial skeleton confirmed that Dart's description of the Taung child was valid for adults as well. There are now hundreds of fossils of A. africanus, largely from the sites of Sterkfontein and Makapansgat in South Africa. Sterkfontein is near Johannesburg and part of the Cradle of Humankind World Heritage Site that includes other hominin sites discussed later in the book. Makapansgat is in the northern part of South Africa. In the late 1990s, a nearly complete skeleton was found at Sterkfontein by Ron Clarke, in a different part of

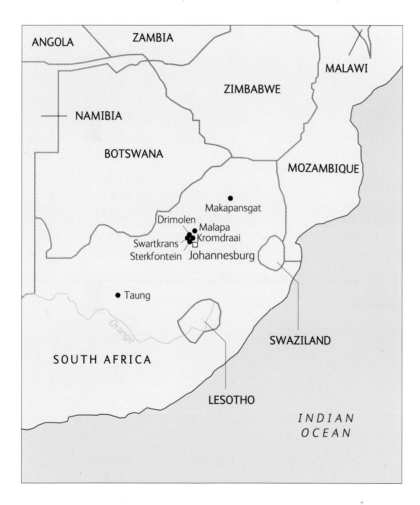

South African sites

The earliest hominins from South Africa are from diverse sites between 1 and just over 3 million years old.

the site from the A. africanus fossils. The difficult excavation and cleaning of this skeleton mean that it has not been fully prepared yet, but it will no doubt tell us much about South African hominins, whether it is found to be A. africanus or to represent an earlier species.

Cave sites

The hominin species described so far have been found at open-air sites such as the extensive and arid badland deposits of eastern Africa and the sandy desert sites in Chad. *Australopithecus* fossils from South Africa were discovered in a completely different context – caves or remnants of caves. Caves differ from open-air sites in how the fossils get into the deposit and how the sites represent time periods. Caves tend to capture specific intervals of

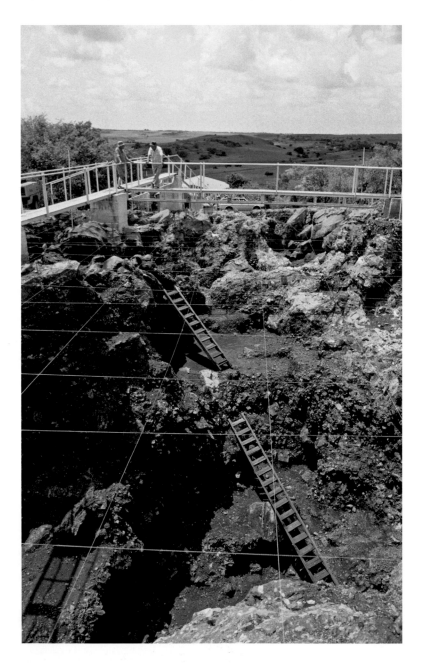

Excavations

Most specimens of *Australopithecus africanus* are from the Sterkfontein site in South Africa, which first revealed hominins in the 1930s. It has now been excavated almost continuously since the late 1960s, producing hundreds of specimens.

time, when they are open for bones, stones and sediment to enter. This material is washed or dropped in, dragged in by carnivores, or it represents animals that have fallen in and become trapped. Later cave sites are sometimes places where humans lived, but this was not the case for a species like *A. africanus*.

Cave geology can be highly irregular, in contrast to the clear horizontal layers that are often seen in open-air sites. Because cave sections can fall or shift over geological time, the general principle that younger sections lay over older sections (the geological 'law of superimposition') does not always apply. Nor are the fossils interspersed among volcanic eruption sequences, as they typically would be in eastern African sites (see pp. 24). The crucial task of dating fossils therefore becomes more complicated.

While sophisticated geochemical methods of dating are increasingly being applied to cave sites, most of our understanding of time comes from the animals themselves. Certain groups of animals are found at different times, and it is possible to develop and refine a timescale based on these cohorts by studying fossil sites with well-established dates. These dates make up a 'biochronology' of how animals have evolved and changed over time, so that when a new site is found, the animals can be compared to the patterns from elsewhere.

Keeping a lookout
Australopithecus africanus may have roamed among the hills near what is now Johannesburg, South Africa.

Based on such comparisons, the *A. africanus* fossils from Sterkfontein and Makapansgat are approximately 2 to 3 million years old. Whether the fossils accumulated over a long period of time or a short one is uncertain, although patterns within Sterkfontein suggest a long period of time. What is clear is that *A. africanus* fossils are found in the absence of stone tools. At least at this time in South Africa, hominins had not yet begun to fashion tools out of stone.

Ecology and diet

The main *A. africanus* sites show a mixture of habitats, from wooded areas to more open bushlands, and there were also rivers nearby. We do not know exactly which habitat would have been preferred by the hominins, or whether *A. africanus* could range successfully through the various environments.

The diet of *A. africanus* has traditionally been contrasted with species of the next section – *Paranthropus*; it did not often eat harder foods. Like *Australopithecus afarensis*, *A. africanus* probably had a broad diet but also the ability to fall back on tougher foods when the opportunity or need presented itself. However, *A. africanus* did have somewhat larger chewing teeth than *A. afarensis*, and the skull morphology indicates a biomechanical rearrangement of muscles, so that force could be applied more efficiently on the back teeth.

Several years ago, researchers began looking at another kind of evidence for diet. Proportions of carbon isotopes in teeth can tell us about diet because different kinds of plants process carbon in different ways. When young animals eat plants, a carbon isotope 'signature' is built into the developing teeth, and this signature is usually resistant to change during millions of years of fossilisation. A downside to this process, as far as the evidence is concerned, is that carnivores have a similar signature to the animals they prey on, so it can be difficult in some cases to distinguish whether hominins were eating plant or animal foods.

The carbon isotope proportions of *A. africanus* teeth suggest that its diet included a significant amount of grass, grass seeds or other animals who themselves ate grass. But there is no

Skull shape

Skulls of *Australopithecus africanus* are an interesting mix of features shared with earlier and later hominins. They had projecting faces like earlier hominins, but cheek bones were moved forward like *Paranthropus*, and the braincase was shaped like *Homo*.

evidence in the teeth that *A. africanus* was specialised for eating grass or grass seeds. Does this mean we should choose the alternative explanation – that *A. africanus* was more carnivorous than we thought? Traditionally, only members of the genus *Homo* have been considered to be meat-eaters.

Chimpanzees are an example of a species that hunts animals for a small part of its food, but does not show the types of adaptations we expect for a carnivorous mammal, such as teeth that can slice meat effectively. In fact, less than 5 percent of the chimp diet is meat. A species like *A. africanus* could have been similar to chimpanzees in this respect, without us being able to see evidence of the behaviour in the bones. But if meat was a major part of the *A. africanus* diet, one would expect some other evidence, either in the shape of teeth or in the presence of stone tools. This question remains a puzzle, and work continues to try to integrate the different lines of evidence for diet.

Sex differences

Forensic anthropologists can determine the sex of a modern human skeleton with 90–95 percent confidence if a pelvis is available, and even have an 80–90 percent success based on skulls alone. Can we do as well with fossil species? Because there are many *A. africanus* specimens compared with most other hominin species, we can investigate sex differences (generally called 'sexual dimorphism') with greater confidence.

Most people know that men have larger browridges than women, and deeper jaws. Such features are mentioned when we talk about 'masculine' and 'feminine' faces. In *A. africanus*, as in humans, the larger individuals tend to have more pronounced browridges. It makes sense to see this as a sex difference, since males are on average larger than females in humans and the apes (this pattern reaches an extreme in gorillas and orang-utans). Because the differences are consistent with sex differences, it would be a mistake to classify large individuals and small individuals into different species. In this way, the patterns of variation we see today in humans and apes help us to understand the fossil record. And, because *A. africanus* is so well known in its skull anatomy, its variation in turn can help us to understand other hominin species.

Life before stone

Even if it used simple tools, *Australopithecus africanus* lived before hominins started to make recognisable stone flakes and choppers.

Australopithecus garhi

Australopithecus garhi was announced in 1999 as a 'surprising' hominin find ('garhi' means 'surprise' in the Afar language spoken in the region where the fossils were found). The fossils are about 2.5 million years old, and the species was defined mainly by one partial skull that possesses very large teeth. A partial skeleton was found at a nearby site, but this did not have a piece of skull, so we cannot be sure that the skeleton and the skull belong to the same species (hominin species are usually defined based on skulls and teeth).

If all of these fossils do represent the same species, then the most interesting feature is the relatively long thighbone. Long legs are a human characteristic not seen in other *Australopithecus* species. The forearms of *A. garhi* are also relatively long, and some have compared the arm proportions to modern orang-utans. This does not mean it is a fossil orang-utan, but simply that it retained an ability to climb effectively. Is that pattern a contradiction with the long legs, usually associated with efficient bipedalism? Only time and further investigation will tell us.

Perhaps more intriguing than the fossils themselves were the cut-marked bones found nearby of the same age of 2.5 million years ago. 'Cut marks' are the signs of stone tools being used to carve up animal carcasses and direct evidence for hominin consumption of meat. The cut-marked bones near *A. garhi* are currently the oldest such evidence, but the research team was careful to recognise that *A. garhi* may not have been the toolmaker and user. It is very difficult to be sure about the association of particular species with the tools found at the same sites. *Australopithecus garhi* may well have more surprises in store.

The skull of *Australopithecus garhi* is broadly similar to other *Australopithecus* species but has very large teeth.

Australopithecus sediba

In 2008, Lee Berger and his then 9-year-old son Matthew visited Malapa, a collapsed cave system not far from Sterkfontein and Swartkrans, and found the first fossils of a new species, *Australopithecus sediba*. The site has yielded the remains of several individuals including the remarkably well-preserved partial skeletons of a boy and an adult female. The sediments surrounding the fossils have been precisely dated to just under 2 million years old using uranium–lead dating and by looking for evidence of reversals in the Earth's magnetic field.

Australopithecus sediba is remarkable for its unexpected combination of primitive and derived characteristics. The small brain and limb proportions are primitive features, similar to those of other Australopithecines. The pelvis of *A. sediba* is also wider than those of other Australopithecines, despite the small brain size. This casts doubt on the idea that the shape of the human pelvis evolved to accommodate larger brained infants during birth. Features in the pelvis suggest that *A. sediba* might have walked upright in a way that was more similar to modern humans than to other Australopithecines. The feet and ankles show a mosaic of primitive and derived features that indicate tree climbing abilities and a unique form of bipedal walking. The hand has relatively short fingers and a long thumb that are characteristic of hands capable of a precision grip, but the muscle attachments also suggest a powerful grasp, useful for climbing trees.

The diet of *A. sediba* included fruit, leaves and bark, something that has not been shown for any other hominin. These food choices suggest a preference for foraging in the forest, which is consistent with aspects of its anatomy, such as long arms and grasping hands that were adapted for climbing trees.

Spectacularly complete fossils such as those from Malapa represent a treasure trove of information. The *A. sediba* fossils reveal that modifications of the pelvis, hand and foot occurred

The skull of an *Australopithecus sediba* child aged between 8 and 13 years old. This skull dates to 1.9 million years ago.

before changes in limb proportions or cranial capacity. The unique mix of primitive and derived features also highlights the difficulty in determining whether a hominin fossil represents a direct human ancestor or a side branch. Some researchers argue that the derived features of *A. sediba* demonstrate a close link to the ancestor of the genus *Homo*, whereas others interpret these features as evidence of homoplasy (meaning that similar features evolved separately on more than one lineage).

Did Australopithecines make stone tools?

Tools are a critical part of human life. As modern humans, we are surrounded by the products of our tools. If, more than 2 million years ago, you observed a group of hominins taking apart an antelope carcass, their use of tools might remind you of your local butcher's shop, even though the hominins looked completely different from humans today.

Evidence for the use and manufacture of stone tools comes from the presence of the tools at a site or from evidence of 'cut marks' on fossil animal bones. Cut marks are the characteristic trace marks left by stone tools when they are used to carve up animal carcasses. They are direct evidence for the use of tools and for the consumption of meat by the hominins who used them. A key question of course is 'Who made and used those tools'. It is very difficult to be sure about the association of particular species with tools or cut-marked bones found at the same site. This is because many sites contain fossils from more than one hominin species, so there may have been other species of hominin present at a site even if their fossils have not been found.

The earliest stone tools come from the site of Gona in the Afar region of Ethiopia and are dated to between 2.6 and 2.5 million years ago. The users of these stone tools have not yet been identified but a possible candidate is *Paranthropus aethiopicus*. The earliest undisputed cut-marked bones are from 2.5 million years ago and were found at the site of Bouri near the fossil of *A. garhi*. Even though these cut marks are clear evidence for the use of stone tools, the research team who found them were careful to recognise that *A. garhi* may not have been the tool maker and user. Making stone tools requires the physical ability to do it effectively.

Adaptations in the hand can reveal whether hominins were capable of making and using tools in a human-like way. We use what is called a precision grip, meaning simply that we can press our thumb and fingertips together with a great deal of strength. This is important for making and using most kinds of stone tools. Although apes and monkeys can grip things between their thumb and fingers, they generate much less force when they do so. The human hand has evolved the ability to grip things precisely and forcefully, so that stones can be rapidly and efficiently used to make sharp flakes and more sophisticated shapes.

When a bonobo (a species of chimpanzee) named Kanzi was trained to make stone tools, he was able to learn this process but then found it easier to simply smash stones on the ground to make sharp flakes. Yet, if making tools is part of your daily life, and crucial to obtaining food, a more efficient method is necessary.

In 2010 researchers described a remarkably complete hand belonging to the partial female skeleton of *A. sediba*. The hand has relatively short fingers and a long thumb, features that are characteristic of hands capable of producing stone tools. So far researchers have not published details of any stone tools from the site of Malapa. Interestingly bark formed a part of the diet of this species and stone tools may have been used to access more edible parts.

Oldowan stone tools

The first stone tools made by hominins are found 2.5 million years ago. They were not sophisticated in today's terms, but to make them still required human adaptations of the hand.

PARANTHROPUS

The *Paranthropus* group is one of the most fascinating branches of the human family tree, largely because they reveal to us a radically different way of being hominin. In a real sense these species were our cousins, not directly on the human line but well within the hominin group. At approximately 2.5 million years ago they diverged from our own lineage and came to be defined by an adaptation to eating hard foods like nuts, seeds and tubers. All of these species are regarded as bipedal, but we are uncertain about the details because not many post cranial (below the skull) skeletons are known for *Paranthropus*. The other name for this group is 'robust *Australopithecus*'. This name is used by scientists who treat them as part of the *Australopithecus* group.

Olduvai Gorge, Tanzania
Fossils found here include species of *Paranthropus* and *Homo*, as well as numerous stone tools.

Paranthropus robustus

What is the hardest thing you can chew? The answers for a human are not very impressive, as we have evolved a lifestyle that involves cooking hard or tough foods so that they are easy to chew and digest. If you were *Paranthropus robustus*, the answer would be completely different. They were able to generate massive forces to crack open hard objects, and signs of this ability make their skulls quite distinctive.

A 'dished' face

A skull of *Paranthropus robustus*, showing the prominent cheekbones that supported powerful chewing muscles.

The first evidence of their remarkable skulls came when *Paranthropus* fossils were found in South Africa in the 1930s. A doctor and palaeontologist named Robert Broom had taken up the search for hominins in South Africa, and his work would prove to be central in defining the hominin fossil record. When describing the face of *P. robustus*, Broom noted an unusual pattern – the cheekbones projected further forward than the nose. If you placed a pencil sideways across the front of the face, it rested on the cheekbones, and the nose would sit back behind the pencil. Broom called their faces 'dished' because the forward projection of the cheekbones made the face resemble a shallow dish.

This sort of appearance is partly because early hominins had relatively flat noses like you would see in a gorilla or chimpanzee. But, more importantly, the cheekbones were brought forward in *P. robustus* to put its chewing muscles in a more effective position. These bones were also very large and robust, and able to support stronger chewing muscles than seen in other hominins. Beneath them was a battery of large premolars and molars – all of the teeth that sit behind the canine and are mainly used for cracking, grinding or slicing foods.

Alongside our ancestors

The name *Paranthropus* means roughly 'equal to' or 'like' humans, and the name *robustus* conveys the impression of large teeth and chewing muscles that has lasted as a general description of this group. The name in this case conveys a real meaning. The species *P. robustus* is not one of our direct ancestors but is simply near us in the evolutionary tree. In many ways, this makes species such

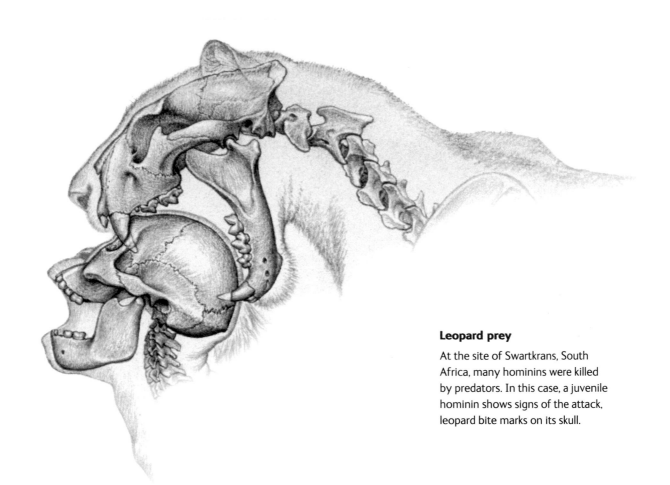

Leopard prey

At the site of Swartkrans, South Africa, many hominins were killed by predators. In this case, a juvenile hominin shows signs of the attack, leopard bite marks on its skull.

Cave sites

Like other hominin sites near Johannesburg, Swartkrans is a former cave. At various times, it was filled from above with sediment, bones, and tools, then later exposed by erosion.

as *P. robustus* more interesting, because it shows that human evolution was not a single path towards modern humans, but instead a more complicated and diverse array of forms. There is more than one way to be a hominin.

The description is also appropriate because fossils of *P. robustus* are usually found at the same sites as our closer relatives, species of *Homo* that are described later. So, if you visited South Africa between 1.5 and 2.0 million years ago, you would see two different kinds of bipeds. One would look relatively familiar as it used tools to obtain meat and vegetable foods. The other – *P. robustus* – would have moved across the landscape consuming mostly vegetable foods, like grasses and sedges, and supplemented these with hard nuts, seeds and tubers that would have been difficult for our own ancestors to chew. *Paranthropus robustus* may also have preyed on small animals for meat, and it probably used simple bone or stone tools. There is also evidence for seasonality in its diet, which might be expected for many early hominins. But, even with a broad diet, *Paranthropus* apparently focused on different resources from *Homo* in order to survive. Both kinds of species were successful in their environments, and indeed many scientists explain their evolution in similar terms. Climates were changing in Africa at the time, producing more open habitats such as grasslands and savannas, and *P. robustus* and *Homo* were adapted to these new environments in different ways.

Some guys just never grow up

Every animal must deal with a number of challenges in life – growing up, getting enough food, avoiding predators, finding a mate and having offspring. Large samples of fossils like those of *P. robustus* actually show us evidence of all of these things. For example, evidence from skulls shows that males of *P. robustus* kept growing long after their wisdom teeth came in, affecting overall size as well as facial appearance. This has implications for social behaviour and the risks they faced during growth.

In humans, males reach maturity later than females, and we describe this by saying that the growth spurt of puberty in boys happens a bit later than in girls. In some other primates, the difference in maturity is much greater than in humans, and by

Size matters

Sexual dimorphism is obvious in this male and female orang-utan.

extending their growth period males can grow to a much larger size than females. For this to make sense in evolutionary terms there should be a benefit in terms of numbers of offspring – essentially a payoff for the time and energy invested in growth. In *P. robustus*, it is likely that the most successful males were those that lived long enough to grow to a large size and outcompete other males, living with a group of females and fathering most of the offspring in that group.

Many males did not survive this long. Adult fossils in the sample of *P. robustus* are biased towards males, and this includes a number of males who were relatively young, with their wisdom teeth recently in. They were apparently killed by predators. When predators and scavengers like leopards and hyenas consume animals, they leave behind marks and characteristic patterns of damage. The Swartkrans site – from which most *P. robustus* fossils come – was originally a cave, so hominin specimens were preserved either because predators consumed them in the cave itself, or perhaps because leopards dragged them up into trees above the cave entrance. This must have happened over and over again, because many individuals of a variety of species are represented, especially hominins.

In the evidence of predation we see the indirect cost of competition among *P. robustus* males. We observe in living primates that males often leave their groups as they reach young adulthood, particularly in those species where competition among males is severe. This results in more time alone, and a particularly dangerous period while they continue to grow to full size. The evolutionary trade-off is that those who make it through this period have a larger number of offspring.

It is possible that this pattern was common among our early ancestors, but at some point in the human lineage the differences between males and females diminished, and it became more common for males to form long-term bonds with females and invest in offspring to a greater extent. In this respect, *Paranthropus* offers another point of contrast to the genus *Homo*, and also to *Ardipithecus*, in which differences between males and females are not so great.

Paranthropus boisei

If there were a ranking of 'extreme hominins', *Paranthropus boisei* would be at the top of the list. While it fits the general description of *Paranthropus robustus*, *P. boisei* takes the same set of adaptations and skull features and exaggerates them. Most scientists view these two species as close relatives, one found in southern Africa and the other found in eastern Africa. The range of *P. boisei* stretches from Ethiopia down to Malawi.

Because of the geological differences between most eastern African sites and the southern African cave sites (see the box on p. 15), we can establish when *P. boisei* was around more accurately than for *P. robustus*. *Paranthropus boisei* lived from 2.3 million years ago until some time after 1.4 million years ago. In other words, it lived in Africa for roughly a million years before it went extinct. By comparison, modern humans have only been around for about 150,000 to 200,000 years.

The first hominin of Olduvai Gorge

The first specimen of *P. boisei* discovered is one of the most dramatic, as it is very well preserved and shows us most of the main aspects of the *P. boisei* skull. It was also the first major hominin find from Olduvai Gorge, the famous site in Tanzania where Mary and Louis Leakey did much of their work. The Leakeys had discovered stone tools at Olduvai, but by the late 1950s there was still no evidence of the maker of these tools. The discovery of *P. boisei* provided a potential toolmaker but perhaps not what had been expected. Instead, like *P. robustus*, it showed more evidence of a different lineage in human evolution. The Leakeys would go on to find evidence of *Homo habilis* and *H. erectus* alongside the *P. boisei* fossils. As in South Africa, it seemed that these two different groups of hominin – *Homo* and *Paranthropus* – lived in the same regions.

The 'Zinj' skull
The skull of *Paranthropus boisei* from Olduvai Gorge, Tanzania, nicknamed after the first scientific name it was given (*Zinjanthropus*).

Koobi Fora, Kenya

Meave and Louise Leakey prospecting in the region of Kenya where abundant fossils of *Paranthropus boisei* are found.

The name *boisei* given to this species does not signify anything about the hominin itself. It was a tribute to the Boise Fund, the organisation that supported the work leading to the discovery of the species. Since it was discovered, *P. boisei* has become central in discussions of human evolution, as large numbers of fossils have been found. These verify the species' distinct anatomy, as seen in the 'dished' face described in the section on *P. robustus*.

My, what small canines you have

Most of us know from television programmes or visits to zoos that monkeys can be very threatening when they open their jaws to reveal large canines. This is especially true of the larger monkeys, such as baboons and mandrills. It is of course a crucial part of their life to be able to make these threats or to use the canines as weapons. In apes, such as chimpanzees and gorillas, the canines are less daunting but still relatively large. In humans, they are small and no longer serve the same function. In *P. boisei*, all of the front teeth were tiny, including the canines.

In a sense, *P. boisei* put all of its tooth-making energy into its back teeth. The difference in size between the tiny, peg-like front teeth (incisors and canines) and the enormous back teeth (premolars and molars) is extraordinary. Although recent results show that *P. boisei* relied heavily on grasses and sedges, their teeth show that they also ate large quantities of hard objects when other foods were scarce, chewing nuts or seeds with their back teeth. This behaviour is distinct from the common way that primates eat, where front teeth are used more extensively to rip flesh off fruits or crop off leaves. It also means that despite its radical adaptations to diet, *P. boisei* had something in common with all other hominins. The human family is defined foremost by having smaller canines than the apes.

Tooth wear

We tend to think of teeth as hard surfaces, and in fact they are the hardest surfaces in our skeleton. If you have always eaten cooked foods and gone to the dentist regularly, the surfaces of your teeth probably look generally the same as they always have, aside from cavities and the dental problems that slowly accumulate. So what would it take to wear the surface away entirely? For other primates, and for most humans in the past, this is a common thing to happen. With a less prepared diet, teeth slowly wear away as they are used to chew foods. The harder or tougher the food, the faster the tooth wears down. At some point, all of the outer layer (the hard enamel) is gone.

This process happened at an amazing rate in *P. boisei*. Despite the fact that *P. boisei* had a thicker enamel layer than any other hominin, the enamel would wear away after just a few years of chewing and grinding up foods. The best illustration of this is the first skull found of *P. boisei*. We can tell from the wisdom teeth that this individual had reached the same point of growth as a human between 15 and 20 years old. (His actual age was younger, because *Australopithecus* and *Paranthropus* species grew up faster than modern humans, but in any case he was not yet adult.) Despite being young, he had already worn away much of the tooth surfaces of the adult teeth that appeared earlier. While tooth wear was pronounced in other hominins, too, groups such as *Australopithecus* do not show this extreme pattern.

A chewing machine
Paranthropus boisei (top) has much larger back teeth than *Australopithecus afarensis* (bottom), but its front teeth are tiny.

Boys and girls

As with *P. robustus*, one of the most intriguing features of *P. boisei* is the variation in size between individuals. Skull size and anatomy were very different between males and females, on a par with the differences seen in gorillas. In gorillas, males are roughly twice the body weight of females. This ratio is much greater than in humans and chimpanzees, in which males tend to be larger than females, but not dramatically so. We can't be sure about body size in *P. boisei*, because little is known of the whole skeleton for this species. However, the skulls show a range of size and shape that allow us to determine the sex of individual specimens. For example, larger skulls have a prominent crest running along the top of the skull, but the smaller individuals do not. This crest is a side effect of hyperdeveloped chewing muscles (specifically the

Different size, same shape

As in gorillas, males of *Paranthropus boisei* were much larger than females.

temporalis muscles). The large muscles reach all the way up from the jaw to the top of the skull, and the extra crest gives them more space to attach.

If this feature helps to identify male skulls, as it does in gorillas, then the larger specimens of *P. boisei* are males, and the smaller ones are females. The degree of size difference leads some to believe that *P. boisei* and *P. robustus* had a similar social system to gorillas, in which large males tend to control groups of females and have most of the offspring while the group is maintained. We can picture small groups of *P. boisei* moving through eastern African woodlands, grasslands and floodplains, with one or two big males and a larger number of females. This is essentially the image of gorillas but in a different, more open landscape. While simplistic, this picture helps illustrate that human evolution involved a variety of different behaviours in different species.

Not just a pretty face

As in *Paranthropus robustus*, the dominant features of the *Paranthropus boisei* face are the massive cheekbones, adapted to a diet of hard foods.

Paranthropus aethiopicus

Perhaps the most evocative nickname in the human fossil record is 'The Black Skull'. That name was given to the discovery in Kenya that was announced in 1985 and essentially defines *Paranthropus aethiopicus*. The Black Skull was so called because it is literally black, but the name is also appropriate because the discovery caused much disagreement among scientists about its place in human evolution.

Based only on *P. robustus* and *P. boisei*, *Paranthropus* would appear to be a relatively uniform group of species, but *P. aethiopicus* is somewhat different from the others and shows an unusual combination of features. Indeed, in the nearly 30 years since the Black Skull was announced, scientists have not been able to reach a consensus about evolutionary relationships among early hominins. They agree about the big picture, but details about species such as *P. aethiopicus* are unresolved.

In some ways, *P. aethiopicus* shows the beginning of typical *Paranthropus* anatomy. It had large chewing muscles and some aspects of the facial appearance of other species in this group. But it also had a more projecting snout, as seen in *Australopithecus* species. The simplest way to see it is as an intermediate species. The original skull dates to 2.5 million years ago, and the few other specimens known for this species are dated to about the same time period, so perhaps this was not a long-lived species compared to others.

To complicate matters, the Black Skull shares some distinctive features with only *P. boisei*, and some scientists would include it in that species. But that would mean that the southern African *Paranthropus*, *P. robustus*, evolved on a completely separate lineage. The similarity between *P. robustus* and *P. boisei* is so pronounced that most scientists think this unlikely. It seems more likely that *P. robustus* and *P. boisei* had a common ancestor that had all of the features they share, but *P. aethiopicus* does not quite fit that description.

A mosaic of evolution

Paranthropus aethiopicus is one of those species that illustrates a pattern of 'mosaic' evolution (see also *Kenyanthropus* on p. 35). This term is used to describe a pattern where features evolve independently and evolve multiple times. A broader example is the separation of human bipedalism and human brain size. We may think of humans as large-brained, bipedal primates, but we became bipeds long before brains reached anything like the size they are now. Human evolution displayed a mosaic pattern instead of a simple progression from one thing to the next. At a more detailed level, unusual discoveries like the Black Skull, which does not fit simple notions of evolutionary paths, also illustrates mosaic evolution.

Unfortunately, the Black Skull all by itself is a large part of what we know about *P. aethiopicus*. Most of the other specimens are fragmentary bits of teeth and jaw, and it is unclear whether the one good skull is typical of its species. *Paranthropus aethiopicus* thus illustrates another common problem in studying the fossil record – sometimes the number of specimens is too small to be sure about the details.

An unusual combination
Nicknamed 'The Black Skull', this specimen from Kenya is 2.5 million years old and is the earliest member of the *Paranthropus* group.

61

EARLY *HOMO*

Of all the names used to describe species in human evolution, *Homo* is the one that means 'human'. It is a familiar term, as modern humans are referred to as *Homo sapiens*, 'intelligent man'. When we speak of early *Homo*, we mean the first signs of our group in particular – species that adopted a way of life we would recognise as human-like. It is contentious who belongs to this group but generally, species found after the appearance of stone tools in the fossil record (2.5 million years ago) and which are close to our line of ancestors are called *Homo*. The three species that appeared relatively early are *Homo habilis*, *H. rudolfensis* and *H. erectus*. Some scientists classify the first two as *Australopithecus*, depending on how they choose to distinguish *Australopithecus* from *Homo*, but in any case, when you see the names '*habilis*' and '*rudolfensis*', they refer to the species described here.

Towards humanity
An elderly, toothless, individual from the *Homo erectus* site of Dmanisi, Georgia – the oldest hominin site outside Africa.

Toolmaker

Homo habilis has long been discussed as the first maker of stone tools, but now a variety of species are found at stone tool sites.

What's for dinner?

Tools serve all kinds of purposes for humans today, but we generally think of stone tools as having one key purpose originally – getting meat and marrow from bones. Since hominins lack the kinds of teeth that carnivores use to slice meat off bone (think of dogs, cats, lions and hyenas), we use tools to do it. You might be surprised how efficiently someone with a few sharp flakes of stone can strip a skeleton of its meat. Early stone tools were also used to break apart limb bones and get at the bone marrow within.

If the evolution of *H. habilis* is linked to stone tools, then it is linked to a change in diet. In addition to the staple foods (such as fruit) that all hominins probably ate when possible, *H. habilis* was able to add larger quantities of meat to its diet. Because harder vegetable foods were less important, the chewing muscles of *H. habilis* were smaller than in *Australopithecus* species and much smaller than in the *Paranthropus* species that lived in the same regions. Early *Homo* tooth surfaces in general also had a slightly more edged or crested shape than in other hominins, and this tooth shape would have helped to chew meat.

When we picture humans eating meat, we may also picture hunting, since that is the way most humans obtain animal meat (other than using domesticated animals, a much more recent phenomenon). Whether or not *H. habilis* and other early *Homo* species were hunters has spurred much debate in human evolutionary research. At some level, hunting is not a unique human feature at all. Chimpanzees hunt occasionally, and they prize meat when they can get it. Perhaps all hominins were killing and eating small animals from time to time. In any case, *H. habilis* was not hunting with the kinds of weapons that human hunters use today, and their key advantage may have been how fast they could get the meat after the animal was dead, rather than how well they could hunt.

Scavenging is often thought of as a possible option for *H. habilis*. This would give quite a different image of the species – hominins that would exploit the kills of larger predators and scavenge meat. Exactly how this would work is unclear, because humans and chimpanzees will usually not eat meat that has been lying around for too long, since it would potentially cause illness. While we do not know the full answer to this question, many scientists would say that *H. habilis* may have hunted small animals and occasionally scavenged when it was able to steal a fresh kill (see also *H. erectus* behaviour on pp. 73–83).

Collections

Archaeologist Fidelis Masao with stone tools from Olduvai Gorge.

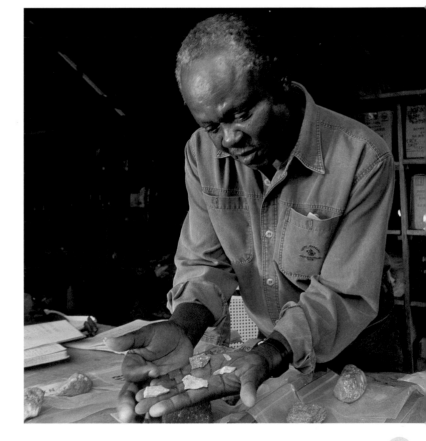

A new diet

Even the most basic stone tools could have transformed the hominin diet to include more meat.

Regardless of how *H. habilis* got its prey, there is another implication of eating meat and bone marrow that affected the evolution of this species. Carnivores tend to have wider home ranges than animals that eat mostly fruits, leaves, nuts and other plant foods. Hunting or scavenging animals successfully usually requires searching over a larger territory. This may be why *H. habilis* is found over a broader area of Africa than *Australopithecus* and *Paranthropus* species.

The little hominin that could

This description may give the impression of a very human-like creature, and indeed that is the heart of the discussion about *H. habilis*. How like us were they? In appearance, *H. habilis* would not stand apart in any dramatic way from their immediate predecessors, the *Australopithecus* species. They were bipeds and the same size as *Australopithecus* – about half the height of a modern human (possibly with larger males, depending on how other fossils are classified – see *Homo rudolfensis*, p. 70). Enormous differences exist between humans today and *H. habilis* then. Microscopic wear patterns on the teeth of *H. habilis* revealed a diet similar to australopithecines, with a preference for fruits and leaves. *Homo habilis* ate a narrower range of foods than later *Homo*, implying that it had not developed the behavioural flexibility to exploit a variety of foods from different environments. Unlike more recent *Homo*, *H. habilis* had strong arms indicating that it spent a lot of time climbing and possibly looking for food in trees. What *H. habilis* shows, once again, is that human evolution was not a single major event. We did not become 'human' all at once.

Homo rudolfensis

Homo rudolfensis is another of the provocative but less well-known species in human evolution. Defining it is a complicated task. For many years after *Homo habilis* was found in the 1960s, more and more fossils were found and classified as *H. habilis*. Eventually scientists became puzzled at some of the patterns of variation in the growing list of specimens. Some things just didn't seem to fit with our idea of how much individuals can vary within a species. For example, the supposed males and females did not show the same kind of differences that we see in modern apes, humans and other hominins. Smaller specimens had more prominent browridges – usually a characteristic of larger males – and the anatomy of the face varied more than it does in the species of today. Studying variation in species is an important task, because many of the biological questions we ask about extinct species – where and when they lived, how they behaved, what caused their evolution and extinction – depend on which specimens are classified to those species.

There is still much debate about *H. rudolfensis*, but basically it represents one solution to the problem of too much variation in *H. habilis*. *Homo rudolfensis* are bigger individuals, with larger brains than *H. habilis* and a facial structure that may indicate differences in how muscles were used to chew food. *Homo rudolfensis* is known from a similar time period to *H. habilis* – the most complete fossils are just less than 2 million years old, while more fragmentary ones are between 2 and 2.5 million years old. The species is named after Lake Rudolf, now called Lake Turkana, in Kenya.

Bigger brains – what do they mean?

One of the more variable features of the broad *H. habilis* sample was brain size. There were individuals with *Australopithecus*-sized brains and other individuals that fell in the range of a more human-like species called *Homo erectus*. When put together, the range of variation struck many people as too much for one species, for example more than you see in chimpanzees or gorillas.

When *H. rudolfensis* is used as a name, it refers partly to the large-brained specimens from Kenya. Does that mean they were smarter than *H. habilis*? Probably not, because the importance of brain size is always relative. Bigger animals have bigger brains just because they are larger on the whole. The skulls of *H. rudolfensis* were also bigger than *H. habilis* in other ways, so relative to overall size, there was nothing special about their brains.

A twist

The problem with describing *H. rudolfensis* as just a big version of *H. habilis* is that they differ in other ways. The face of the most important *H. rudolfensis* specimen is flatter from side to side, and its cheekbone is further forward. Its flatter face looks more like *Australopithecus* than you might expect for a skull that is otherwise more human-like. What this means is unclear, but if *H. rudolfensis* and *H. habilis* were eating different foods, or differed ecologically in other ways, it would help explain why they were able to live in the same region without competing too much for the same resources.

Splitting species

Homo rudolfensis (left) and *Homo habilis* (right). Differences between these specimens lead many scientists to recognize two different species.

An alternative explanation of *H. rudolfensis* is that it simply is not necessary to recognise different species, or that the current classification is not the right solution to variation in the early *Homo* group; some scientists continue to interpret the '*H. rudolfensis*' fossils as part of normal variation in a broadly defined species that would be *H. habilis*. But recently described fossils from Koobi Fora are a further indication that there was more than one species of early *Homo* in East Africa nearly 2 million years ago.

Earliest *Homo*?

Who wins the prize as the oldest specimen of *Homo*? As you can imagine, this is important to know if we want to explain the evolution of *Homo* based on the relationship to another event, such as the first stone tools. The oldest stone tools discovered so far are dated to just over 2.5 million years ago from the Gona site in Ethiopia, and this is an accurate date based on the process of radioactive decay (see p. 15).

Fossil hominins, however, are rarer at 2.5 million years ago (bones do not preserve as well as stone), and they belong to *Australopithecus* or *Paranthropus* species. The best candidates for the oldest member of *Homo* are an upper jaw from Ethiopia (clearly *Homo*, just over 2.3 million years ago), a lower jaw from Malawi (probably *Homo*, possibly as old as 2.4 million years, but the date is less certain), and a piece of a skull from Kenya that mainly contains bones around the ear opening (maybe not *Homo* at all, but the date is over 2.4 million years). Depending on who you ask, these fossils may be attributed to *H. habilis*, *H. rudolfensis* or *Homo* in a general sense without a species name.

An upper jaw of the genus *Homo*, found at Hadar, Ethiopia, and dated to just over 2.3 million years ago.

It is frustrating not to know the exact answer to this question and to have only a few specimens from this time period. Another problem is that much of what we know about *H. habilis* and *H. rudolfensis* comes from the more complete fossils just under 2 million years old. The earlier fossils may belong to these species, but scientists have not yet confirmed whether they have all of the features we are interested in (e.g. brain size or features of hand bones). If you interpret the record as it stands, stone tools were first made and used by hominins prior to the origin of the genus *Homo*, but we need a better record between 2 and 3 million years ago to resolve the order of events and find out who the first toolmakers truly were.

Homo erectus

What defines success in the evolution of a species? Is it overall population size? Geographic range? Impact on the world? Or is it simply how long the species survives? If the latter is the case, then we have a clear winner in human evolution, and it is not us. *Homo erectus* lived from almost 2 million to roughly 200,000 years ago. Depending on what is called *H. erectus*, the latest sites in Java could be as recent as 50,000 years old. Over nearly a 2-million-year span of time we find evidence of more or less the same species. Some other hominins were successful in this sense, too – *Paranthropus boisei* survived approximately 1 million years, and the exact spans of earlier species may have been similar. But the lifespan of *H. erectus* as a species is truly impressive for the hominins we know well. Although 'success' is a fuzzy term in biology, it is clear that something about the *H. erectus* way of life enabled it to survive a period that included many swings in climate and environment.

Indeed, *Homo erectus* is often described in terms of a series of achievements – they were the first hominins to spread out of Africa, in many places they used a new type of tool (the handaxe), and they were probably the first hominins to control fire. They were perhaps the first hominins who were not merely skilful in the sense of *H. habilis* making tools, but also able to adapt as humans do. It is always wrong to describe evolution in directed terms, as if *H. erectus* was 'trying' to be human. They were not. They were very good at being *H. erectus*, and that was enough.

Out of Africa, part one

All hominins before *H. erectus* lived only in Africa. This is true for other early members of the genus *Homo*, *H. habilis* and *H. rudolfensis*. But almost as soon as *H. erectus* appears, it is found in Africa, western Asia, eastern Asia and the southeast Asian islands of Indonesia. Within a rapid span of time, this species spread further than other hominins had done. Why did this happen?

Animals expand their ranges for lots of reasons. It is worth remembering, of course, that *H. erectus* individuals would not have known where they were going. They were essentially seeking

Handaxe
A modern replica of a handaxe, along with the debris created when the stone tool is made.

Spreading out

Homo erectus sites are spread over much of Africa, Asia and possibly Europe. In some places, indicated by the question marks, sites with stone tools are known from more than a million years ago and were probably made by *Homo erectus*, but there are no hominin fossils that allow us to be sure.

out new places to get food and to have less competition while doing it. And even though 100,000 years is a short time in geological terms, spreading across Asia over this period would not have required much migration in the life of a single person. This was not an epic journey of a single group or population, but instead the sort of gradual expansion of range that one sees in other animals, too.

The reason why hominins had previously all remained in Africa was probably largely because their habitats did not stretch across the Sinai region connecting Africa and Asia. Today, this is a hot, dry area, and it is a narrow stretch of land. *Homo erectus* almost certainly did not cross this region under conditions anything like today, but at times in the past the Sinai would have looked different, as local climate fluctuated from more arid to more humid and lush. We can think of groups of *H. erectus* drifting across northern Africa when grasslands, woodlands and prey animals stretched that far, and some groups would have found themselves crossing into Asia without realising the significance humans would later attach to it.

This event had something to do with how *H. erectus* was living, as well. We know that some hominins had started to make meat more important in their diet even back at 2.5 million years ago. Carnivorous animals such as lions often range more widely than herbivorous animals, because their prey are mobile and have less predictable locations than plants do. Indeed, lions themselves once had a range including parts of Asia and Europe, as well as Africa. *Homo erectus* fits the carnivore pattern, and it was also larger than previous hominin species. An increase in human size, combined with a shift to meat-eating, would explain a wider geographic range, a range that extended out of Africa.

Improvements in technology

It is clear that *H. erectus* did not need new kinds of stone tools to survive at first. The tools found at the earliest *H. erectus* sites are the same kinds seen with earlier species, and *H. erectus* was using this technology when it first spread out of Africa. During the entire lifespan of the *H. erectus* species, however, some things did change. Handaxes are found in Africa from about 1.7 million years ago. These were larger tools that required more preparation than the simple choppers that existed before, but they are not found everywhere that *H. erectus* is.

What we do not know is when exactly *H. erectus* started doing other things differently. The use of fire is perhaps the most important innovation during the era of *H. erectus*, as hominins who controlled fire would have been able to cook food. Cooking is important not just for meat but also to make tough vegetables easier to eat. Vegetable foods – whether raw or cooked – are just as important as meat for human populations living in tropical climates.

Fire

Many scientists think *Homo erectus* was the first hominin to control fire, though the exact date of this major event is difficult to determine.

There are hints of the use of fire at early *H. erectus* sites, but more commonly accepted evidence of fire does not come until later. For example, *H. erectus* lived in the area of Swartkrans, a site in South Africa that also has fossils of *Paranthropus robustus*. There are burnt bones at Swartkrans that resemble bones cooked for some time in a fire rather than bones exposed to a quick brush fire. It is critical to find ways to separate natural fire from human-controlled fire, and the evidence at Swartkrans is the earliest point where this may be possible. The burnt bones at Swartkrans are between 1.0 and 1.5 million years old, close to the origin of *H. erectus*, so it is possible that the use of fire was always a part of the life of *H. erectus*. Cooking would allow easier consumption of both meat and vegetable foods, and it may well be the reason why human chewing muscles are dramatically reduced compared to those of our ancestors.

Life in Asia

Although we describe *H. erectus* spreading out from Africa to Asia, the discovery of fossils happened in reverse. *Homo erectus* has always been known as an Asian species, from the first discoveries by Eugene Dubois at the Trinil site in Java, Indonesia, in the late nineteenth century. And it is in Asia that *H. erectus* survived for so long. In Africa, *H. erectus* evolved into species described in the next section of this book, such as *Homo heidelbergensis*. In Asia, *H. erectus* remained much the same as it was initially. Scientists often refer to Asian *H. erectus* as 'classic *Homo erectus*', since these discoveries defined the species.

Asian *H. erectus* fossils are very robust – the bones of the skull are thick, and features such as browridges and other ridges on the skull are pronounced. The brain size was less than that of modern humans but greater than *H. habilis*, *Australopithecus* or *Paranthropus*. These 'robust' features are not the same as the robust features used to describe *Paranthropus* species (see pp. 48–61).

Sangiran, Java, Indonesia

Some of the first discoveries of *Homo erectus* in Asia were made here.

A classic Asian species

A *Homo erectus* cranium from the Sangiran site in Java, Indonesia, showing the exaggerated browridge and thick skull bones.

China

One of the localities at the Zhoukoudian *Homo erectus* site in China. *Homo erectus* survived here long after *Homo heidelbergensis* evolved in Africa.

For reasons to do with their physiology, growth patterns and overall level of activity, *H. erectus* individuals had thick bones, but the shape of the skull, for example, was broadly human-like rather than adapted to a diet requiring massive chewing muscles.

Some of the youngest evidence of *H. erectus* – 300,000 to 500,000 years in age – comes from an extensive set of specimens from Zhoukoudian in China, 50 kilometres southwest of Beijing. This is a famous site, and the first finds in the 1920s and 1930s there gave rise to the nickname 'Peking Man'. Sadly, many of the best fossils were lost during the Second World War, and only casts remain. Work still continues at Zhoukoudian, providing new discoveries.

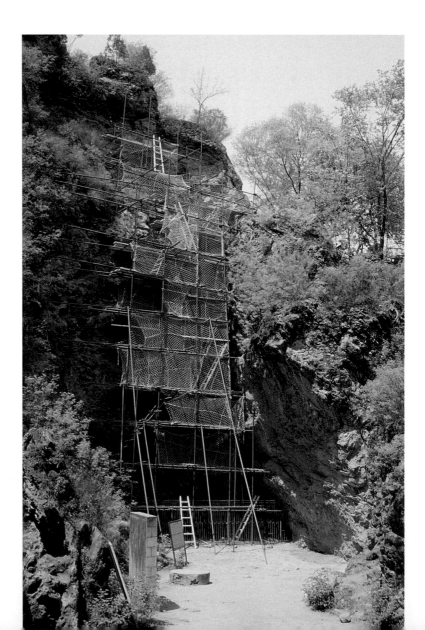

Investigating the scene

If you want to know how species such as *H. erectus* and *H. habilis* behaved, you have to look for a certain kind of forensic evidence. Using tools to take apart a skeleton and scrape off the meat leaves a characteristic pattern of damage and marks. The marks are 'cut marks', literally the marks left by stone flakes when bones are cut apart or scraped. Such marks are common on animal bones found at places like Olduvai Gorge, and there is something remarkable about looking at the work of a butcher from nearly 2 million years ago. The patterns on the bones indicate that *H. erectus* (and probably *H. habilis*, too) was able to get at the best joints of meat. Marks are found, for example, close to the hip and shoulder joints of the animals they ate (a variety of antelope species and occasionally some larger mammals

Behaviour

Homo erectus showed many similar behaviours to modern humans, though the extent of human-like language in this species is unknown.

Complete skeleton

The 'Turkana Boy' skeleton from Nariokotome, Kenya, is the most complete individual of *Homo erectus*.

such as hippos). If *H. erectus* had been a scavenger, these thicker portions of meat would have been long gone by the time it had its turn. Instead, the pattern of marks suggests that *H. erectus* was either hunting the animals itself or immediately stealing kills from other predators, as modern human hunters sometimes do. Either way, this meat butchering and eating was a significant part of its behaviour.

A growing boy

The discovery that has most influenced our understanding of *H. erectus* in recent times is the Nariokotome boy. Nariokotome is a site in the Lake Turkana region of Kenya that features frequently throughout the human story. In 1984, a team led by anthropologists Richard Leakey and Alan Walker discovered one of the most complete skeletons that has been described for any hominin before more recent species such as the Neanderthals. The Nariokotome boy (hip anatomy shows that he was a male) was only at the stage of growth of a 12-year-old human and may, in fact, have been only about 9 years old, given a faster rate of development. Yet he was 160 cm tall (5 foot 3 inches). This height is similar to tall humans of that age and taller than earlier hominins. His skeleton is also human-like, and he is thought to have walked much as we do.

Unlike early hominins such as Lucy (see pp. 22–30), *H. erectus* probably did not use trees for safety or as a main source of food. Climbing was most likely a rare activity, as in modern humans – useful sometimes for hunting or gathering foods, but not the critical way of getting around that it was for apes and early hominins. We think *H. erectus* lived in a range of habitats, including more open habitats like modern savannas and grasslands. They may have been safer in these habitats than earlier hominins, because they were larger or if they were able to control fire. The Nariokotome boy gives us an image of *H. erectus* as a tall, human-like species, walking across the landscape as humans do today.

Another very human feature of Nariokotome is that he had skeletal problems. In particular, he suffered from curvature of the spine (scoliosis) and had a hunched posture. His own individual story is just as interesting as the implications for *H. erectus* as a species.

A more complicated pattern

Discoveries in recent years have revealed more variation in
H. erectus, showing an unexpected and more interesting pattern.
The site of Dmanisi, in the Republic of Georgia in western Asia,
astounded scientists through the 1990s and early twenty-first
century. They discovered a series of well-preserved skulls that
looked like *H. erectus* and yet did not show all of the expected
features. In particular, some of them have small brain sizes similar
to the range found in *H. habilis*. The skulls from Dmanisi are all
around 1.8 million years old, which is at the beginning of the time
when *H. erectus* lived. This makes us wonder how different the
H. erectus populations were from each other, and it also shows
that the first hominins to leave Africa were more similar to
H. habilis than we thought. Further discoveries may even show
that *H. habilis* itself spread out of Africa.

First stop

The Dmanisi early hominin site in the
republic of Georgia was discovered
beneath the ruins of a medieval city.

'Homo ergaster'

'*Homo ergaster*' is the name sometimes given to early African specimens of *H. erectus*, and it is sometimes used for the newer discoveries from Dmanisi in Georgia as well. The name 'ergaster' comes from the word 'workman' in its Greek form, and it was named this way for reasons similar to the naming of *H. habilis*. The African fossils are found with stone tools, and in some cases these are more advanced tools than those used by *H. habilis*.

Early African fossils of *H. erectus* do not have all of the 'classic' *H. erectus* features associated with finds first made in Asia. The simplest description is that they are less robust, and this overall effect is seen in a number of correlated features of the skull. Scientists who separate '*H. ergaster*' from *H. erectus* generally think that the African

Some scientists refer to African *Homo erectus* fossils as a different species, '*Homo ergaster*', because they do not have all of the features seen in Asian populations of *Homo erectus*.

species is on the human line, and that the Asian specimens of *H. erectus* eventually went extinct rather than evolving into other forms. The impressive Nariokotome skeleton would belong to '*H. ergaster*' if this name was used.

Regardless of how populations are named, after about 0.8 million years ago *H. erectus* was no longer found in Africa. Although *H. erectus* survived until relatively recently in Asia, the African populations became something different, the subject of the next section.

Differences also exist between the African and eastern Asian specimens of *H. erectus*, and for this reason early African specimens are sometimes called '*Homo ergaster*' (see box on p. 82). When we study a hominin that lived on several continents, we are faced with complex evolutionary questions that do not apply to earlier hominins. *Homo erectus* reveals the challenge of studying the past – the more we know, the more questions we ask, and the more limited the evidence seems. Despite being relatively well known, and surviving such a long time, *H. erectus* is still full of surprises.

Discovery in context
One of the skulls from Dmanisi before it was removed in 2001.

LATER *HOMO*

After one million years ago, we see hominin species beyond
Africa, living across Asia and in parts of Europe. At this point
in human evolution, we think of these hominins as distinctly
human. To some scientists, all of the species described here as
'later *Homo*' belong to our own species, *Homo sapiens*. If that
is the case, they are all populations within an evolving species,
one that changes over time and shows some regional variations
but is essentially one genetically cohesive group. A few scientists
would go as far as including *Homo erectus* in *H. sapiens* as well.
This book adopts the more common view – that the recent fossil
record of human evolution includes a variety of species. One
implication of recognising species diversity is that not all of the
species are in the direct line of human evolution. The modern
human lineage is just one branch of the hominin tree.

Old England
Excavation at the 500,000-year-old site of Boxgrove, one of the most important
sources of evidence for early hominin life in Europe.

Homo heidelbergensis

The name 'Homo heidelbergensis' has been around since it was first coined for a jaw found near Heidelberg, Germany, in 1907. But the concept of H. heidelbergensis is essentially a new and changeable idea. It is most easily thought of as a descendant of Homo erectus in Africa, which spread out to Europe. In this way H. heidelbergensis was the ancestor for Neanderthals and modern humans.

If we assume that only one species existed in Africa and Europe between 300,000 and 700,000 years ago, H. heidelbergensis is the name of that species. It is recognisable from its human-like but very robust and massive skull. Some individuals had brain sizes falling in the modern human range of 1200–1500 cubic centimetres (about three times as big as Australopithecus brains). Skulls of H. heidelbergensis from places like Greece, Ethiopia and Zambia are remarkably similar to each other, which is why the species is thought of as both European and African. As time passes, however, European populations begin to look different, giving rise to the well-known Neanderthals.

Human arrival in Britain

A flint tool found at Pakefield, one of the earliest hominin sites in Britain. We do not know the human species responsible (see p. 90).

The 'archaic' face

Homo heidelbergensis individuals lacked chins and had much larger browridges than modern humans. (A reconstruction of Bodo from Ethiopia.)

First specimen

The name *Homo heidelbergensis* is based on this lower jaw from the Mauer site, near Heidelberg, Germany.

The oldest spears

Scientists often debate the evolution of hunting behaviour. This is relevant to technology and archaeology, to cooperation among individuals, to what the hominins ate and to how far they ranged. The *H. heidelbergensis* site of Boxgrove, England suggests that big game hunting, perhaps using wooden spears, was already in place 500,000 years ago. Actual wooden spears are known from about 400,000 years ago in England, and slightly later from a site in Germany. They are the oldest recognisable weapons (tools used for hunting rather than butchering) in the human fossil record, and horses found in the same site were probably the targets.

As extraordinary as they are, these spears also remind us of what we do not see in the fossil record. Wooden tools are only preserved in the rarest of circumstances – the archaeological record is almost entirely focused on stone and bone. The wooden spears in Germany are relatively simple, sharpened poles, without stone tips. Weapons like these might have existed in much older times, without ever being preserved. In any case, they tell us that these early Europeans were hunting large animals and in this sense were human-like.

LATER *HOMO*

87

Hunter

The Atapuerca site in Spain has a vast number of fossil specimens, some of which are compiled here as a representative skeleton of an early Neanderthal soon after the evolutionary transition from *Homo heidelbergensis*. These early Neanderthals also used wooden spears, as the oldest ones known are from the same time period.

What is an 'archaic human'?

Homo heidelbergensis is known mostly from skulls, but the few known postcranial bones give the impression of a human-sized, heavily built, bipedal hominin. Muscle markings are more pronounced than in most modern humans, and limb bones have thick layers of hard bone around their central marrow cavities. Reconstructions often portray these people with 'bodybuilder' physiques. These aspects of size are important. Humans are the second biggest modern primates, after gorillas. A lot of resources are needed to maintain such size, and *H. heidelbergensis* faced this challenge as much as modern humans.

Lacking chins, and with a longer, lower shape of the skull, they appear 'archaic', retaining some characteristics of our earlier ancestors. This appearance, rather than brain size, is why the term 'archaic humans' is sometimes used to describe *H. heidelbergensis* and some other recent species. Their behaviour was broadly human-like, but if you were to encounter a *H. heidelbergensis* person, you would see them as a different species, especially because their facial appearance was not within the normal, modern human range of variation.

Homo heidelbergensis lived during a time when multiple hominin species lived on Earth. It is easy to slip into thinking of human evolution as a linear process, from one species to the next until modern humans evolved. But when *H. heidelbergensis* evolved in Africa and Europe, *H. erectus* continued successfully in eastern Asia (read the section on *H. erectus* to see just how long that species survived). For hundreds of thousands of years there were at least two human-like hominins living at the same time but in different geographic regions. Only one of them gave rise to later humans, and this, we think, was *H. heidelbergensis*.

Before Neanderthals

A *Homo heidelbergensis* skull from Petralona, Greece, showing the basic human shape but more robustly built.

Climate change

The warm and hospitable climate of northern Europe during *H. heidelbergensis* times was not permanent. This is obvious from the present climate, but even hundreds of thousands of years ago there were periods when Europe became much colder than it is now. During the Pleistocene – a geological term for the epoch stretching from 1.8 million years ago up to 10,000 years ago – the Earth had begun its swings between ice ages and warm stages, or, more formally, between glacial and interglacial phases. As time went on, the swings became more and more extreme. The changes in climate may have forced *H. heidelbergensis* to migrate across the continent to track its prey and to avoid the harshest areas.

The cold periods may also have affected the anatomy of *H. heidelbergensis* in Europe, because natural selection favoured individuals better able to survive the low temperatures. This phenomenon is well illustrated by Neanderthals, the species that followed *H. heidelbergensis* in Europe. As the most recent European specimens we call *H. heidelbergensis* are different in some ways from the earlier specimens, how to divide the two groups is unclear. In a sense, the last *H. heidelbergensis* in Europe were also the first Neanderthals.

Homo antecessor

Some time before 700,000 years ago, a group of hominins found themselves on a warm subtropical coast, with long summer nights and short winter days. They were surrounded by other animals adapted to warm climates, such as elephants and hippos. Those hominins did not know that they were living in what would eventually be called England, known for quite different weather. At that time, the world was in a warm phase, and many tropical animals – including our ancestors – had spread north as far as northern Europe. The site of Pakefield on the eastern English coast provides one of the oldest examples of humans living in the British Isles, but the evidence consists only of their tools. The species that made the tools is a bit of a mystery, but it may have been a *Homo* species known from Spain at about this date, *Homo antecessor* ('Pioneer Man').

The period between early *Homo erectus* and modern humans may seem to be a simple pattern, with *Homo heidelbergensis* filling the role of a 'transitional' species. But a glance through descriptions of human evolution will reveal a variety of names that suggest multiple lineages at this time. All of these names come down to the question of what species are and how much difference between populations is enough to mark them as different species. For the period between 800,000 and 1 million years ago, another hominin name that you might see is *Homo antecessor*.

Homo antecessor is a more complicated story, based on the oldest fossil hominins found in Europe. The mountainous northern Spanish site of Atapuerca is famous not only for its remarkable collection of hominin fossils from about 400,000 years ago, but also for the discovery of European fossils, dated to 800,000 years. The most informative specimens are mainly parts of the skull of a teenager. They were called *H. antecessor* because of differences in facial anatomy between them and other hominins – especially the presence of a 'sunken' area below the cheekbone called the canine fossa, also a human feature. Not many fossils are known from this time, so it is unclear how many species there were. Some scientists have called other early Europeans *H. antecessor* as well, but for simplicity many refer to all of them as *H. heidelbergensis*. Until there is solid agreement on the details of human evolution in the last million years, the names may change somewhat.

It's all in the face
Homo antecessor shows some facial differences from *Homo heidelbergensis*.

LATER *HOMO*

Revolutionary find

The first recognised specimen of a Neanderthal, discovered with parts of a skeleton in the Neander Valley, Germany, in 1856.

Homo neanderthalensis

When you hear 'Neanderthal', you probably think of it as a broad term for human ancestors. In reality, *Homo neanderthalensis* is the best-known hominin species from the past and specifically a European and west Asian group. Ironically, it is also one of the greatest mysteries. We know an extraordinary amount about the Neanderthals' behaviour, where they lived, and when, but scientists still debate whether they are an extinct hominin species or a population that is part of the modern human lineage. The image of a Neanderthal is familiar – a projecting face, no chin, a large browridge, and a human-sized brain in a skull that was longer from front to back than a human skull. But what do these features mean? How different were Neanderthals from humans today?

Brawn and brains?

Some have described Neanderthals as using brawn more than brains to solve their problems. As with *H. heidelbergensis*, muscle markings and bone thickness indicate a powerful physique. This physique was normal for most of human evolutionary history. No doubt Neanderthals were strong people, but this does not mean they had limited intelligence. In fact, their brains were as large as a modern person's brain. In a broad sense, Neanderthal behaviour was not that different from other hominins living at the same time in other parts of the world. They hunted, and they may have hunted more than other populations. Europe was cold during most of Neanderthal existence, and in cold climates humans rely more on animals for food. Neanderthals had also mastered the control of fire.

What is particularly interesting about Neanderthals is the amount of damage visible in their skeletons. Despite their strength and their robust bones, Neanderthals show high rates of broken bones and dislocated joints, along with dental diseases and arthritis. The bone damage and arthritis show the amount of stress they put on their skeletons. Neanderthals were not 'superhuman' in any sense. It is likely that many injuries came from hunting large animals at close range – Neanderthals preyed on bison and horses as well as smaller animals like red deer and gazelles. Neanderthals used spears but did not have technology such as bows and arrows or spear-throwers to help them kill their prey from a long distance.

In many respects, Neanderthals were like other hominins from the last several hundred thousand years, the group of hominins sometimes called 'archaic humans'. Modern humans today are less muscular and have less robust bones than our ancestors. This is true even for humans living in traditional ways, such as hunting and gathering foods. For Neanderthals, the question is whether they were a distinct species for other reasons.

It's cold up here

When human ancestors first arrived in Europe, it was semi-tropical – a logical place to go if you were a species that evolved in hot Africa. But, over time, Europe went through 'glacial' periods of extreme cold. The changing climate caused movement of animals back and forth to follow their habitats, but in many cases it also produced evolutionary change. Animals that were more highly adapted to cold conditions were better able to survive. These animals included reindeer, a frequent target of Neanderthal hunting during cold periods. As part of a process of evolution, European hominins transformed from *H. heidelbergensis* to the species we recognise as Neanderthals.

Neanderthal tools

Neanderthals used new, more efficient techniques to produce stone tools referred to as 'Middle Palaeolithic' or 'Mousterian' tools. These were critical in carving meat from bison, reindeer and other large mammals.

LATER *HOMO*

93

Many mammals show a pattern of adaptation to temperature. Mammals living in higher latitudes (further north or south, in places with colder winters) are better at conserving heat, while mammals closer to the equator are better at losing heat and staying cool when the weather is extremely hot.

How did this affect Neanderthals? We describe Neanderthals as short and stocky, and their relatively short arms and legs are considered to be an adaptation to cold environments. Heat loss in humans is affected by how much surface area the body has, in other words how much skin is exposed to the weather. With short arms and legs, humans become better at keeping heat in. Modern humans that live in the far north, such as some European populations and especially the Inuit of North America and the Sámi populations of northern Eurasia, tend to be different from other humans in this way. Neanderthals show a more extreme form of this adaptation. That said, Neanderthals lived in a broad swathe of territory from western Europe all the way east to what is now southern Siberia, and south to sites in Israel. It appears that their range may have shifted with climate, and that they were not capable of handling the most severe temperatures.

Burying the dead

In trying to decide how human-like the Neanderthals were, we need to consider what their social behaviour was like and how their groups were structured. In some ways they were strikingly like us. For example, Neanderthals buried their dead, whether for reasons of honouring them, or to maintain hygiene in the places in which they lived. One of the most famous and debated finds is the discovery of flower pollen with a Neanderthal burial at the Shanidar site in Iraq. It could indicate the symbolic burial of flowers with a person, or the pollen may have got there by other means, such as flowers taken by rodents down their burrows. Some of the most recent Neanderthals may also have made simple jewellery for themselves, such as necklaces of pierced animal teeth. Were they 'modern' because of this?

The differences between Neanderthals and modern humans have to do with quality and quantity. Early in their history, modern humans used similar tools to Neanderthals, but later in time they

Another burial

An infant Neanderthal skeleton from Amud, Israel. Even at a young age (about 10 months old), we can see Neanderthal features such as the long foramen magnum visible here, and the shape of the lower jaw.

made more sophisticated and more diverse tools, showing more delicate techniques (see *Homo sapiens*, pp. 102–109). Modern humans used beads and ornaments earlier than Neanderthals (some scientists conclude that Neanderthals learned these techniques from modern humans), and they eventually expanded their artwork to include cave paintings and a variety of sculpted objects. Either there were simply fewer Neanderthals around – which would make sense in the cold European habitat – or they had less need and interest in the kinds of sophisticated art that modern humans produced. However, a recent discovery at two sites in Spain of evidence for body painting, suggests that these differences might be less than we previously thought. As the presence of pigment-stained shells is widely accepted as evidence of modern humans having symbolic thinking, we may now need to re-examine the evidence for symbolism in the Neanderthals.

The first Neanderthals

Asking who was the first Neanderthal is like asking who was the first human. Because evolution is a process of change, it is impossible to agree on the point where species become what they are. In a sense, we can 'see' Neanderthals evolving by the appearance of some Neanderthal features at the extraordinary site of Atapuerca, northern Spain, about 400,000 years ago. For many years, scientists also referred to specimens like these as 'pre-Neanderthals' because of their intermediate anatomy.

By 100,000 years ago, Neanderthals possessed the many distinct features that allow us to recognise them, including large, projecting faces and large noses; a braincase that was human-sized but much lower and longer from front to back; and robust skeletons with curved limb bones. They also have features on the back and underside of the skull that are very rare in modern humans, such as a projecting mound of bone on the back of the skull called an occipital bun. Individuals with all of these characteristics are often called 'classic' Neanderthals.

Neanderthal physique
Neanderthals were shorter and stockier than most humans today.

hair type to skull shape and brain functions. The latest results support the view that the Neanderthals separated from our lineage about 400,000 years ago. But when the new Neanderthal genome was compared in detail with those of living people from different regions, the results produced an intriguing twist to our evolutionary story – the genomes of people from Europe, China and New Guinea lay slightly closer to the Neanderthal sequence than did those of Africans. Thus, if you are European, Asian or New Guinean, you probably have about 2% of Neanderthal DNA in your genetic make-up.

Evolutionary trends

The original Neanderthal skullcap (bottom row, middle) in a line of skulls belonging to – from upper left – *Australopithecus africanus*, *Homo rudolfensis*, *H. erectus*, *H. heidelbergensis*, and, on the bottom right, a modern human (*H. sapiens*). Although it is possible to line up skulls like this through time, the actual pattern of human evolution is more complex.

A possible explanation for this discovery is that the ancestors of today's Europeans, Asians and New Guineans interbred with Neanderthals in western Asia as they exited Africa about 60,000 years ago. That ancient human exodus may have involved only a few thousand people, so it would have taken the absorption of only a few Neanderthals into a group of *H. sapiens* for the genetic effect – greatly magnified as modern human numbers exploded – to be widespread tens of thousands of years later.

The breakthrough in reconstructing a Neanderthal genome has been mirrored across Asia in work on an ancient human group that has become known as the 'Denisovans'. A fossil finger bone about 50,000 years old found in Denisova Cave, Siberia, was too incomplete to be assigned to a human species, but its whole genome has now been reconstructed. This has revealed a previously unrecognised Asian offshoot of the Neanderthal line, but one that has also impacted modern humans. These Denisovans are related to living humans in only one region – Australasia (Australia, New Guinea and neighbouring regions). As well as a small Neanderthal component, these modern humans also carry about 3% of Denisovan DNA from another interbreeding event that must have happened as their ancestors passed though South East Asia about 50,000 years ago. Thus the big picture of our evolution is that our species is of recent African origin (more than 90% of our genetic ancestry), but in detail there are relics of earlier interbreeding events with other kinds of human still remaining in our DNA.

Telling a story

There has been much debate about whether Neanderthals used language. It is likely that they did, even if the way they spoke was different from modern humans.

LATER *HOMO*

99

Flores site

The site of the excavation of *Homo floresiensis* in the Liang Bua cave on the island of Flores, Indonesia.

Homo floresiensis

Homo floresiensis is the most recent discovery in the genus *Homo*. If it is indeed a distinct species, it could also have been the last of our close relatives to have gone extinct. This surprising hominin was isolated on the island of Flores in Indonesia and apparently lived there less than 50,000 years ago. It immediately became known as 'the Hobbit' because of its diminutive stature, but this gives a wrong impression of its appearance – the fictional hobbits have large brains, unlike the Flores hominins! It is true, however, that the Flores people were only about 1–1.2 metres (3–4 feet) tall.

Since being announced in 2004, the Flores fossils have aroused much controversy and debate. Their brain sizes were roughly the same as apes, even after accounting for their small body size. The rest of their anatomy shows that they belong in the human family, but their skeletons do not exactly match those of any other species. The Flores population is thought to be an example of 'island dwarfing', a process that happens to many large mammals when they become isolated on islands. Indeed, alongside the Flores fossils tiny elephants were found that were a source of food for *H. floresiensis*.

Other explanations?

The main hypothesis explaining Flores is that this group started as an isolated population of *Homo erectus* and eventually evolved to a small size with tiny brains. A number of individuals have been found, and they are all small, but only one complete skull is known, and some scientists believe that he or she suffered from microcephaly, a genetic disorder where the brain does not grow normally but remains small throughout life. If this is so, then brain size in the population as a whole may have been bigger, and the question of whether this group is a distinct species is more complicated. Some argue that both the small stature as well as the brain size can be explained by pathologies that occur in modern humans, and that the Flores fossils are not a distinct species after all.

The Flores population used tools like other populations of the genus *Homo*. Were their brains somehow reorganised to allow them to have tiny brains but still make tools? Some researchers have found differences in brain shape that support this conclusion. Brains are costly organs in terms of the energy they require, and on a small island hominins may have evolved to be more energetically efficient by having smaller brains.

Other aspects of the Flores skeleton are also distinct from modern *Homo sapiens*. Scientists have now studied bones of the arms and legs, and hands and feet, and in each case found patterns of anatomy that are outside the range of variation seen in humans today. In some ways, the Flores skeleton is more similar to *H. habilis*, *H. erectus* or *Australopithecus*, or indeed unique among hominins. How much of this unusual anatomy is due to evolution in an island environment with constrained resources? Are some of these differences the effect of disease or genetic abnormalities? The Flores discoveries have revealed a number of things we do not know about the recent human fossil record. The number of specimens of *H. erectus*, for example, is very limited except for skulls, and this makes it more difficult to interpret Flores. It may even be that the Flores hominin evolved from an earlier hominin species, whose skeleton we know comparatively little about.

Hobbit-sized

The original skull and jaw of *Homo floresiensis*, with the lower jaw of a different individual of similar size.

LATER *HOMO*

101

Homo sapiens

What does it mean to be human? There are many answers to this question in human evolution. For an introduction like this one, we can ask what it means to be *Homo sapiens*. The answer to that question defines the beginning of our species in the past, as well as what we are today. It is not an easy task. Because the human fossil record is complete enough to show us a variety of human-like groups in the last million years, there is no single point where the most significant transition took place.

Some would take that to mean that *H. sapiens* is a variable species with a long history. If so, it would include several of the species described in this book, such as Neanderthals. But it is more common to think of *H. sapiens* in specific anatomical terms, as we do with other species. When one does that, it turns out that modern humans have a recent origin.

African origin

Reconstruction of a *Homo sapiens* skull discovered in 1967 in the Omo region of southern Ethiopia. The skull is now dated at nearly 200,000 years old.

An African origin

Humans evolved in Africa. The very first hominins known lived in Africa, and the best evidence for the first modern humans is found there too. In the fossil record, modern humans are referred to specifically as 'anatomically modern *Homo sapiens*'. They are people whose fossil remains are within the range of variation in modern human body shape, or close to it. This phrase does not apply to the Neanderthals, who are anatomically different from us.

The first skulls of *H. sapiens* in this sense are from Ethiopia, from sites between 150,000 and 200,000 years ago. That does not neccessarily mean that modern humans evolved in Ethiopia first. Where fossils are found depends largely on geology and where it is possible to find them. Modern humans are also found soon afterwards in other parts of Africa, such as South Africa.

The oldest human skulls are recognisable from their chins, the high arching foreheads, the generally rounded shape of the skull, and the characteristic human face. Modern human skulls have a slightly 'sunken' appearance in the face, especially the bones just below the cheekbones. This gives us our prominent cheeks.

Such a portrait is of course just an anatomical description. It is like a forensic anthropologist identifying a person, but we are identifying a species. In their behaviour, the earliest modern humans were not much different from other hominin groups at the time. Their tools were similar, and presumably their modes of hunting, gathering resources and structuring their social groups were similar as well. When we say 'anatomically modern', we only mean how they are recognised.

Works of art

When then did humans begin to behave the way we do? Many would say that, to understand modern human evolution, behaviour is more important than the anatomy. They suggest we should seek out evidence of humans *thinking* like they do today. This behaviour could express itself as symbolic behaviour, art, burying artefacts with the dead – anything that shows people imagining their place in the world and looking beyond material existence.

Before Picasso

Early modern humans in Europe left behind spectacular cave art, as seen in this panel from Lascaux, France.

The most dramatic early works of art by *H. sapiens* are the cave painting sites in France and Spain, and 'Venus' figurines found in Germany, Austria, France, Czech Republic and Russia. The most ancient of these are about 35,000 years old and close to the arrival of modern humans in this region. The people that made these paintings and sculptures are called 'Upper Palaeolithic', the term used to describe their technology. In a broader sense, the Upper Palaeolithic period began only about 45,000 years ago, long after the earliest modern humans are found in the fossil record. It may be that humans were essentially 'human' long before they showed clear evidence of symbolic behaviour. In southern Africa and Australia, similar artwork is found from about 30,000 years ago as well.

On the other hand, such flashy works of art may only be the flourish that came after a long period of this behaviour. More simple ornaments such as beads, and evidence of ochre being used as body paint, go back to similar dates as the first anatomically modern humans. According to this evidence, humans adopted modern behaviour at about the same time as they became anatomically modern human and in the same place – Africa.

Where and when

Not long after they appeared in Africa, modern humans spread to the Middle East. Their bones are found there at sites older than 100,000 years ago. This venture out of Africa was brief – climate change seems to have pushed modern humans back into Africa after 100,000 years ago, and Neanderthals lived in the Middle East for some time after modern humans did. Then, around 50,000–60,000 years ago, modern human fossils appear again in the Middle East and soon afterwards as far away as Australia and other parts of Asia. This event is the second major migration out of Africa for hominins. Modern humans spread out to other regions, replacing other hominin populations or occupying lands left empty after those hominins became extinct.

Who were the other hominins? *Homo neanderthalensis, H. floresiensis* and possibly *H. erectus* were still around after modern humans evolved in Africa, and were living in Europe and Asia. We do not have a clear picture of all these groups, mainly because the fossil record of Asia is poor during the key time period, although recent genetic analysis of a finger bone found in Siberia, shows that a population sharing common ancestry with Neanderthals was in Asia at this time (see pp. 96–99). Neanderthals are known relatively well, and they and modern humans lived in different places for a long time. But something seems to have happened 50,000–60,000 years ago to allow modern humans to spread out successfully. We do not know what

Out of Africa again

A skull of *Homo sapiens* from the Skhul site in Israel, more than 100,000 years old.

Modern human skulls

A selection of skulls of *Homo sapiens*. The two skulls in the middle are among the earliest anatomically modern humans known from Europe (left) and China (right), and the two outer skulls are modern people from the same regions. Recent skulls are often less robust because of changes in lifestyle.

this was. It may be that modern human technology made them more successful in direct competition with other hominins, but it may also be that other hominins were not coping as well with climate change and went extinct. Another possibility is that language took its modern form at this time, and it allowed the communication of more complex ideas within *H. sapiens* but not in other species.

By 30,000 years ago, modern humans had replaced Neanderthals in Europe, and all other hominin species had probably gone extinct. By 15,000 years ago, modern humans had spread to North and South America. Now, we are essentially in the modern era of the evolutionary record. *Homo sapiens* is unique in having such a broad geographical range, compared to previous hominins or to other mammals. Evidence of a massive expansion of geographical range was seen previously in *H. erectus*, but modern humans truly spread worldwide.

Cro-Magnon man

A skull from the Cro-Magnon site in France. Discovered in the nineteenth century, this site gave its name to the first modern humans in Europe.

Why?

Humans did not spread throughout the world in order to conquer it. We can still describe these events in evolutionary terms, because population pressure encourages animals of all kinds to spread out when possible. What is different in this case is the wide range of environments occupied by humans and that they replaced other hominins in many places. To a large extent, modern humans were probably more efficient at doing the same things that other hominin species did. In many respects, early modern humans had the same lifestyle as Neanderthals and *Homo heidelbergensis* – hunting animals for meat and gathering a variety of vegetable foods. They were largely nomadic or moved with the seasons. Domestication of animals, the development of agriculture and more settled communities were a much later occurrence. We do not know whether modern humans competed directly with other hominin populations, but greater efficiency in obtaining food and more sophisticated cultural adaptations would explain the rapid spread of modern humans through the world, and our eventual occupation of more extreme habitats than other hominin species.

Along the way, humans developed a seemingly endless variety of behaviours, cultures and languages. We cannot describe in simple terms what humans eat, how they obtain their food and how they interact with one another as we did for the earlier species in this book, such as *Australopithecus*. It is true that we can describe humans in strictly biological terms, such as bipedal, omnivorous with an emphasis on meat and other cooked food, with complex social structure and relatively large body size, but this description does not capture what we are. Human intelligence interacts with the material world to enable us to live in an extraordinary variety of different ways, through our tools clothes, and shelter. This great capacity for variation and adaptation is what makes humans human.

Overview – the big picture

After reading this book, you may be bewildered by the number of names used for human ancestors or relatives. These names are essentially the result of pushing the fossil record to the greatest

Human dispersal

The first appearance of *Homo sapiens* in the Middle East may have been temporary, as Neanderthals occupied the region after the first modern humans. At about 50,000–60,000 years ago, *Homo sapiens* spread out of Africa again, replacing other hominin populations.

The last hominin, so far...

An anatomically modern human in Europe, from the era of cave paintings and symbolic objects.

limit of detail. But the difficulty with species names is that they are usually coined when new discoveries are made and when the least information is available about a new group. Only over time do we settle on an agreed way of categorising information about these groups, and it is safe to say that the study of human evolution is still changeable in this sense.

It is important to step away from the names and see the broad pattern, because then we see the remarkable way that the fossil record tells the story of human evolution. It reveals any number of so-called intermediate forms – not ape and not human – which we expect to see in the process of evolution. It shows that the array of features we call 'human' evolved at different times. Human ancestors were bipeds long before brain size increased or stone tools were made. Finally, even the most conservative taxonomist sees diversity of hominin species in the past, especially in the discovery of *Paranthropus* hominins living in the same regions as our direct ancestors. All of these conclusions stem from fossil discoveries, and they represent a consensus that is generally agreed while the finer points of human evolution are debated.

The next step?

Even though we can look back and identify the first signs of human anatomy and behaviour, and describe a process of human evolution, that process was not pre-ordained. It also does not allow us to predict what will happen next. With species such as *Homo erectus*, *Paranthropus boisei*, and others, we see how successful some species were for long periods of time. They were not simply transitional groups waiting to evolve into something else. This is true of *Homo sapiens* as well. Anatomically, we are much the same as we were 100,000 years ago. Behaviourally, on the other hand, modern humans show an incredible range of diversity, and technology now evolves much faster than biology. In that shift from biological to technological evolution, the topics of this book give way to recent archaeology and modern history – different subjects but with the same goal of explaining the human past. What comes next is unknown, but we will see the effects of technological change and our own decisions long before humans become something we would call a different species.

Index

Italic pagination refers to illustrations.

Further information

Further reading

Boyd, R., and J. Silk. *How Humans Evolved*. 5th ed. Norton, 2009.

Cartmill, M., and F. Smith. *The Human Lineage*. 2nd ed. Wiley-Blackwell, 2009.

Fagan, B. *Cro-Magnon: How the Ice Age Gave Birth to the First Modern Humans*. Bloomsbury Press, 2010.

Johanson, D., and K. Wong. *Lucy's Legacy: The Quest for Human Origins*. Harmony Books, 2009.

Klein, R. G. *The Human Career*. University of Chicago Press, 2009.

Potts, R., and C. Sloan. *What Does It Mean to Be Human?* National Geographic, 2010.

Reader, J. *Missing Links*. Oxford University Press, 2011.

Stringer, C. *The Origin of Our Species*. Penguin, 2011.

Stringer, C., and P. Andrews. *The Complete World of Human Evolution*. Thames & Hudson, 2011.

Tattersall, I. *Masters of the Planet: The Search for Our Human Origins*. Palgrave Macmillan, 2012.

Wood, B. *Human Evolution (a Brief Insight)*. Sterling, 2011.

Zimmer, C. *Smithsonian Intimate Guide to Human Origins*. Harper, 2007.

Websites

Website addresses are subject to change. There are many websites that deal with human evolution, but here are a few key sites for your interest.

American Museum of Natural History: http://www.amnh.org/

BBC: http://www.bbc.co.uk/sn/prehistoric_life/human/human_evolution/

Center for Academic Research and Training in Anthropogeny: http://www.ucsd.tv/carta/

Institute of Human Origins: http://www.becominghuman.org

Leakey Foundation: http://www.leakeyfoundation.org

Natural History Museum: http://www.nhm.ac.uk/

New Scientist: http://www.newscientist.com/channel/being-human/human-evolution

Smithsonian Institution: http://www.mnh.si.edu/anthro/humanorigins/

The Talk Origins Archive: http://www.talkorigins.org/faqs/homs/

Picture Credits

p.6 © John Sibbick/NHMPL; p.8 © MPFT; p.10 Ben Plumridge © Thames & Hudson Ltd, London, from The Complete World of Human Evolution (2005) p.185; p.11t © Gerald Eck; p.11b © Institute of Human Origins. Photo: William Kimbel; p.12 © MPFT; p.13t,m Mike Eaton © NHMPL; p.13b © Ann Gibbons; p.14 Mike Eaton © NHMPL; p.15 © Charles Lockwood; p.16 © Sileshi Semaw; p.17 National Museum of Ethiopia, Addis Ababa. © 1994 Tim D. White\Brill Atlanta; p.18 © NHMPL; p.20 © Charles Lockwood; p.22t,m © C. Owen Lovejoy; p.22b © Charles Lockwood; p.23l © John Sibbick/NHMPL; p.23r © NHMPL; p.24 Lisa Wilson © NHMPL; p.25 © John Reader/Science Photo Library; p.27 © Jay H. Matternes 1984; p.28 © Charles Lockwood; p.29 NHMPL; p.30 © Zeresenay Alemseged; p.31 © Koobi Fora Research Project/Meave Leakey; p.32 © National Museums of Kenya; p.33 © W. Schnaubelt N. Kieser-Atelier WILD LIFE ART for the Hessische Landesmuseum Darmstadt (HLMD); p.35 © National Museums of Kenya. Photo: Fred Spoor; p.36 © Don Johanson, Institute of Human Origins; p.37 Lisa Wilson © NHMPL; p.38 © Institute of Human Origins. Photo: William Kimbel; p.39 © John Sibbick/NHMPL; p.41 © Mauricio Anton/Science Photo Library; p.42 © Jay H. Matternes 2003; p.44 National Museum of Ethiopia, Addis Ababa. Photo © 1999 David L. Brill\Brill Atlanta; p.45 Courtesy University of the Witwatersrand, Johannesburg. Photo: Brett Eloff; p.47 © NHMPL; p.48 © Neil Paskin; p.50 © Charles Lockwood; p.51t © Jay H. Matternes 1984; p.52b © Chris Stringer; p.52 © Institute of Human Origins. Photo: William Kimbel; p.53 ©2012 Visionquest Entertainment International. Photographer: Holly Carroll; p.55 © NHMPL; p.56 © Fred Spoor; p.57t © Don Johanson, Institute of Human Origins; p.57b © Gerald Eck; p.58t © Don Johanson, Institute of Human Origins; p.58b © Yoel Rak; p.59 © Jay H. Matternes 1999; p.61 © Human Origins Program, Smithsonian Institution; p.62 © Kenneth Garrett; p.64 © NHMPL; p.65 © NHMPL; p.66 © Jay H. Matternes 1999; p.67 © Kenneth Garrett/National Geographic Image Collection; pp.68–69 © John Sibbick; p.71 © NHMPL; p.72 © Institute of Human Origins. Photo: William Kimbel; p.73 © John Reader/Science Photo Library; p.74 Lisa Wilson © NHMPL; p.76 © Jay H. Matternes 1982; p.77 © Chris Turney. Bones, Rocks and Stars: The Science of When Things Happened. Published by Macmillan Science, June 2006; p.78t © NHMPL; p.78b © Paul Goldberg; p.79 © Mauricio Anton/ Science Photo Library; p.80 © National Museums of Kenya; p.81 © Georgian National Museum. Photo: Guram Tsibakhashvili; p.82 © John Holmes/NHMPL; p.83 © Georgian National Museum. Photo: Guram Tsibakhashvili; p.84 © The Boxgrove Project; p.86t © AHOB/NHMPL; p.86b © Mauricio Anton/ Science Photo Library; p.87 © NHMPL; p.89 © NHMPL; p.90l © Mauricio Anton/Science Photo Library; p.90b © Javier Trueba/MSF/ Science Photo Library; p.91r © John Sibbick/NHMPL; p.91l © Javier Trueba/MSF/Science Photo Library; p.92 © NHMPL; p.93 © NHMPL; p.94 Yoel Rak; p.95 © John Sibbick/NHMPL; p.96t © NHMPL; p.96b Courtesy João Zilhão; p.97 Lisa Wilson © NHMPL; p.98 © NHMPL; p.99 © Jay H. Matternes 1982; p.100 © Fletcher & Baylis; p.101 © Peter Brown; p.102 © M. Day/NHMPL; p.104 © Pascal Goetgheluck/ Science Photo Library; p.105 © NHMPL; p.106 © John Reader/ Science Photo Library; p.107 Lisa Wilson © NHMPL; p.108 © John Sibbick/NHMPL.

Every effort has been made to contact and accurately credit all copyright holders. If we have been unsuccessful, we apologise and welcome correction for future editions and reprints.

(AHOB, Ancient Human Occupation of Britain project; MPFT, Mission Paleoanthropologique Franco-Tchadienne; NHMPL, Natural History Museum Picture Library; MSF, Madrid Scientific Films)